Everyday Mathematics®

The University of Chicago School Mathematics Project

Home Links

Grade **2**

McGraw Hill **Wright Group**

The McGraw·Hill Companies

The University of Chicago School Mathematics Project (UCSMP)

Max Bell, Director, UCSMP Elementary Materials Component; Director, *Everyday Mathematics* First Edition; James McBride, Director, *Everyday Mathematics* Second Edition; Andy Isaacs, Director, *Everyday Mathematics* Third Edition; Amy Dillard, Associate Director, *Everyday Mathematics* Third Edition

Authors

Max Bell, Jean Bell, John Bretzlauf, Amy Dillard, Robert Hartfield, Andy Isaacs, James McBride, Cheryl G. Moran*, Kathleen Pitvorec, Peter Saecker

*Third Edition only

Technical Art

Diana Barrie

Teachers in Residence

Kathleen Clark, Patti Satz

Editorial Assistant

John Wray

Contributors

Librada Acosta, Carol Arkin, Robert Balfanz, Sharlean Brooks, Jean Callahan, Ann Coglianese, Mary Ellen Dairyko, Tresea Felder, James Flanders, Dorothy Freedman, Rita Gronbach, Deborah Arron Leslie, William D. Pattison, LaDonna Pitts, Danette Riehle, Marie Schilling, Robert Strang, Sadako Tengan, Therese Wasik, Leeann Wille, Michael Wilson

Photo Credits

Front cover (l)Linda Lewis/Frank Lane Picture Agency/CORBIS, (r)C Squared Studios/Photodisc/Getty Images, (bkgd)Estelle Klawitter/CORBIS; **Back cover** Estelle Klawitter/CORBIS; **iii iv** The McGraw-Hill Companies; **v** Image Club; **1** (t)Photodisc/Getty Images, (bl)The McGraw-Hill Companies, (br)Scott Gibson/CORBIS; **2** (t)Mazer Creative Services, (c bl br)The McGraw-Hill Companies; **3** (t)The McGraw-Hill Companies, (b)Photodisc/Getty Images; **11 29** The McGraw-Hill Companies; **30** (chair)Peter Dazeley/Photographer's Choice/Getty Images, (television)Ryan McVay/Getty Images, (egg)Siede Preis/Getty Images, glasses)Stockdisc/PunchStock, (all others)The McGraw-Hill Companies; **43** (tl tc tr)The McGraw-Hill Companies, (br)Image Source/Alamy; **85** Getty Images; **87** (tl tr c)The McGraw-Hill Companies, (bl)Gary Conner/PhotoEdit, (br)Jeremy Woodhouse/Digital Vision/Getty Images; **108** (l)John A. Rizzo/Getty Images, (r)Ingram Publishing/Alamy; **131** The McGraw-Hill Companies; **132** Tom & Dee Ann McCarthy/CORBIS; **149** The McGraw-Hill Companies; **150** Mark Karrass/CORBIS; **151** John A. Rizzo/Photodisc/Getty Images; **157** The McGraw-Hill Companies; **167** (l)Ingram Publishing/Fotosearch, (r)Comstock Images/Getty Images; **169** (t)Brand X Pictures/Getty Images, (b)MedioImages/CORBIS; **195 197 222 224 225 227** The McGraw-Hill Companies; **245** Photodisc/Getty Images; **265** Ingram Publishing/Fotosearch.

---

www.WrightGroup.com

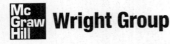

Wright Group

Printed in the United States of America.

Send all inquiries to:
Wright Group/McGraw-Hill
P.O. Box 812960
Chicago, IL 60681

ISBN-13 978-0-07-609739-5
ISBN-10 0-07-609739-0

17 18 19 HES 14 13 12 11

The McGraw·Hill Companies

Contents

Unit 1

1•1   Unit 1: Family Letter 1
1•11 Relations: <, >, = ...5
1•12 Temperatures 7
1•13 Unit 2: Family Letter 9

Unit 2

2•1   Addition Number Stories 13
2•2   Addition Facts 15
2•3   Doubles Facts 17
2•4   Turn-Around, Doubles, and +9 . . .19
2•5   Addition Facts Maze 21
2•5   Domino Facts 23
2•7   Fact Triangles 25
2•3   Weighing Things 29
2•9   Name-Collection Boxes. 31
2•10 Frames-and-Arrows Problems . . .33
2•11 "What's My Rule?". 35
2•12 Subtraction Maze 37
2•13 Addition/Subtraction Facts 39
2•14 Unit 3: Family Letter 43

Unit 3

3•1   Place Value. 47
3•2   How Much Does It Cost? 49
3•3   Times of Day. 51
3•4   "What's My Rule?" with Blocks . . .53
3•5   Pockets Bar Graph 55
3•6   Frames-and-Arrows Problems . . .57
3•7   Change at a Garage Sale 59
3•8   Counting Up to Make Change. . . .61
3•9   Unit 4: Family Letter 63

Unit 4

4•1   Change Number Stories 67
4•2   Parts-and-Total Number Stories . .69
4•3   Reading a Thermometer. 71
4•4   Temperature 73
4•5   Shopping at the Grocery Store . . .75
4•6   Addition Number Stories 77
4•7   Measuring to the Nearest Inch . . .79
4•8   Addition Strategies 81
4•9   Place Value. 83
4•10 Unit 5: Family Letter 85

Unit 5

5•1	"What's My Attribute Rule?"	89
5•2	Line Segments	91
5•3	Parallel Line Segments	93
5•4	Polygons	95
5•5	Quadrangles	97
5•6	3-D Shapes	99
5•7	Make a Triangular Pyramid	101
5•8	Symmetry Hunt	103
5•9	Unit 6: Family Letter	105

Unit 6

6•1	Adding Three Numbers	109
6•2	Comparison Number Stories	111
6•3	Graphing Data	113
6•4	Number Stories and Diagrams	115
6•5	Subtracting with Base-10 Blocks	117
6•6	How Many?	119
6•7	How Many?	121
6•8	Arrays	123
6•9	Arrays	125
6•10	Division	127
6•11	Unit 7: Family Letter	129

Unit 7

7•1	Count by 2s, 5s, and 10s	133
7•2	Missing Addends	135
7•3	Who Scored More Points?	137
7•4	Doubles and Halves	139
7•5	Estimating Weights	141
7•6	Comparing Arm Spans	143
7•7	Find the Middle Value	145
7•8	Interpreting Data	147
7•9	Unit 8: Family Letter	149

Unit 8

8•1	Equal Parts	153
8•2	Fractions of Shapes	155
8•3	Fractions of Collections	157
8•4	Shading Fractional Parts	159
8•5	Fractions of Regions	161
8•6	More or Less Than $\frac{1}{2}$?	163
8•7	Fractions	165
8•8	Unit 9: Family Letter	167

Unit 9

9-1	Using Measurement	171
9-2	Linear Measurements	173
9-3	Measuring Lengths	177
9-4	Perimeter	181
9-5	Travel Interview	185
9-6	Capacity and Area	187
9-7	Area and Perimeter	189
9-8	Capacity	191
9-9	Weight	193
9-10	Unit 10: Family Letter	195

Unit 10

10-1	Coin Combinations	199
10-2	How Much?	201
10-3	Coin Values	203
10-4	Calculators and Money	205
10-5	Estimation to the Nearest 10¢	207
10-6	Making Change	209
10-7	Area	211
10-8	Place Value	213
10-9	Counting by 10s, 100s, and 1,000s	215
10-10	4-Digit and 5-Digit Numbers	217
10-11	Grouping with Parentheses	219
10-12	Unit 11: Family Letter	221

Unit 11

11-1	Buying Art Supplies	225
11-2	Comparing Costs	227
11-3	Trade-First Subtraction	229
11-4	Multiplication Stories	231
11-5	Division Number Stories	233
11-6	Multiplication Facts	235
11-7	Multiplication Facts	237
11-8	$\times, \div$ Fact Triangles	239
11-9	Fact Families	243
11-10	Unit 12: Family Letter	245

Unit 12

12-1	Fact Triangles	249
12-2	Many Names for Times	251
12-3	Timelines	253
12-4	$\times, \div$ Fact Triangles	255
12-5	$\times, \div$ Facts Practice	259
12-6	Typical Life Spans	261
12-7	Interpret a Bar Graph	263
12-8	Family Letter	265

HOME LINK
1·1

Unit 1: Family Letter

Introduction to *Second Grade Everyday Mathematics*

Welcome to *Second Grade Everyday Mathematics*. It is a part of an elementary school mathematics curriculum developed by the University of Chicago School Mathematics Project.

Several features of the program are described below to help familiarize you with the structure and expectations of *Everyday Mathematics*.

A problem-solving approach based on everyday situations
By making connections between their own knowledge and their experiences both in school and outside of school, children learn basic math skills in meaningful contexts so the mathematics becomes "real."

Frequent practice of basic skills Instead of practice presented in a single, tedious drill format, children practice basic skills in a variety of more engaging ways. Children will complete daily review exercises covering a variety of topics, find patterns on the number grid, work with addition and subtraction fact families in different formats, and play games that are specifically designed to develop basic skills.

An instructional approach that revisits concepts regularly
To improve the development of basic skills and concepts, children regularly revisit previously learned concepts and repeatedly practice skills encountered earlier. The lessons are designed to build on concepts and skills throughout the year instead of treating them as isolated bits of knowledge.

A curriculum that explores mathematical content beyond basic arithmetic Mathematics standards around the world indicate that basic arithmetic skills are only the beginning of the mathematical knowledge children will need as they develop critical-thinking skills. In addition to basic arithmetic, *Everyday Mathematics* develops concepts and skills in the following topics—number and numeration; operations and computation; data and chance; geometry; measurement and reference frames; and patterns, functions, and algebra.

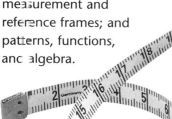

Second Grade Everyday Mathematics emphasizes the following content:

Number and Numeration Counting; reading and writing numbers; identifying place value; comparing numbers; working with fractions; using money to develop place value and decimal concepts

Operations and Computation Recalling addition and subtraction facts; exploring fact families (related addition and subtraction facts, such as $2 + 5 = 7$, $5 + 2 = 7$, $7 - 5 = 2$, and $7 - 2 = 5$); adding and subtracting with tens and hundreds; beginning multiplication and division; exchanging money amounts

Data and Chance Collecting, organizing, and interpreting data using tables, charts, and graphs

Geometry Exploring and naming 2- and 3-dimensional shapes

Measurement Using tools to measure length, weight, capacity, and volume; using U.S. customary and metric measurement units, such as feet, centimeters, ounces, and grams

Reference Frames Using clocks, calendars, thermometers, and number lines

Patterns, Functions, and Algebra Exploring number patterns, rules for number sequences, relations between numbers, and attributes

Everyday Mathematics provides you with many opportunities to monitor your child's progress and to participate in your child's mathematics experiences.

Throughout the year, you will receive Family Letters to keep you informed of the mathematical content that your child will be studying in each unit. Each letter includes a vocabulary list, suggested Do-Anytime Activities for you and your child, and an answer guide to selected Home Link (homework) activities.

You will enjoy seeing your child's confidence and comprehension soar as he or she connects mathematics to everyday life.

We look forward to an exciting year!

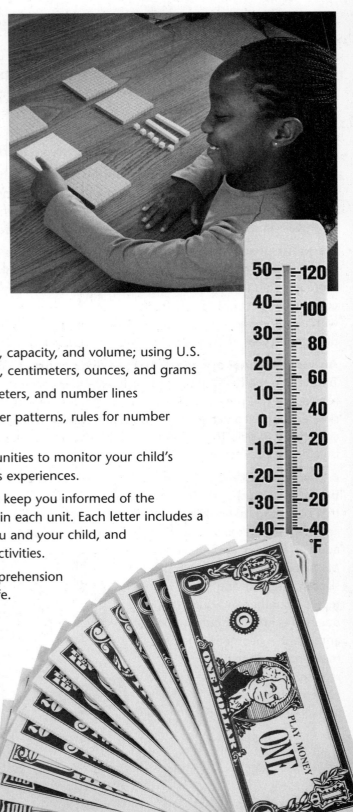

2

Unit 1: Numbers and Routines

This unit reacquaints children with the daily routines of *Everyday Mathematics*. Children also review and extend mathematical concepts that were developed in *Kindergarten Everyday Mathematics* and *First Grade Everyday Mathematics*.

In Unit 1, children will ...

◆ Count in several different intervals—forward by 2s from 300, forward by 10s from 64, backward by 10s from 116, and so on.

◆ Practice addition facts, such as $5 + 4 = ?$ and $? = 7 + 5$.

◆ Review whole numbers by answering questions like "Which number comes after 57? After 98? After 234?" and "Which number is 10 more than 34? 67? 89?"

◆ Respond to prompts like "Write 38. Circle the digit in the 10s place. Put an X on the digit in the 1s place."

◆ Work with a number grid to reinforce place-value skills and observe number patterns.

-9	-8	-7	-6	-5	-4	-3	-2	-1	⓪
1	2	③	4	5	⑥	7	8	⑨	10
11	⑫	13	14	⑮	16	17	⑱	19	20
㉑	22	23	㉔	25	26	㉗	28	29	㉚

←Children use number grids to learn about ones and tens digits and to identify number patterns, such as multiples of three.

◆ Review equivalent number names, such as $10 = 5 + 5$, $10 = 7 + 3$, $10 = 20 - 10$, and so on.

◆ Play games, such as *Addition Top-It*, to strengthen number skills.

◆ Practice telling time and using a calendar.

Do-Anytime Activities

To work with your child on the concepts taught in this unit, try these interesting and rewarding activities:

1. Discuss examples of mathematics in everyday life: television listings, road signs, money, recipe measurements, time, and so on.

2. Discuss rules for working with a partner or in a group.
   - ◆ Speak quietly.   ◆ Be polite.   ◆ Help each other.
   - ◆ Share.   ◆ Listen to your partner.
   - ◆ Take turns.   ◆ Praise your partner.
   - ◆ Talk about problems.

3. Discuss household tools that can be used to measure things or help solve mathematical problems.

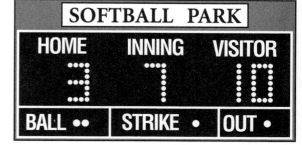

Vocabulary

Important terms in Unit 1:

math journal A book used by each child; it contains examples, instructions, and problems, as well as space to record answers and observations.

tool kits Individual zippered bags or boxes used in the classroom; they contain a variety of items, such as rulers, play money, and number cards, to help children understand mathematical ideas.

Math Message A daily activity children complete independently, usually as a lead-in to the day's lesson. For example: "Count by 10s. Count as high as you can in 1 minute. Write down the number you get to."

Mental Math and Reflexes A daily whole-class oral or written activity, often emphasizing computation done mentally.

number grid A table in which numbers are arranged consecutively, usually in rows of ten. A move from one number to the next within a row is a change of 1; a move from one number to the next within a column is a change of 10.

-9	-8	-7	-6	-5	-4	-3	-2	-1	0
1	2	3	4	5	6	7	8	9	10
11	12	13	14	15	16	17	18	19	20
21	22	23	24	25	26	27	28	29	30

Exploration A small-group, hands-on activity designed to introduce or extend a topic.

Math Boxes Math problems in the math journal that provide opportunities for reviewing and practicing previously introduced skills.

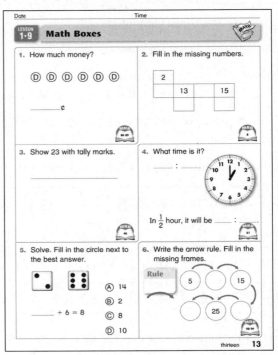

Home Links Problems and activities intended to promote follow-up and enrichment at home.

As You Help Your Child with Homework

As your child brings home assignments, you may want to go over the instructions together, clarifying them as necessary. The answers listed below will guide you through this unit's Home Links.

Home Link 1·11

1. <   2. >
3. >   4. =
5. Answers vary.   6. Answers vary.

Home Link 1·12

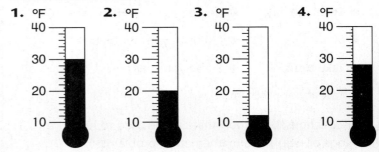

HOME LINK 1·13

Unit 2: Family Letter

Addition and Subtraction Facts

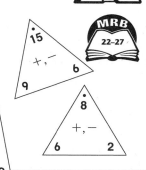

Unit 2 focuses on reviewing and extending addition facts and linking subtraction to addition. Children will solve basic addition and subtraction facts through real-life stories.

In *Everyday Mathematics*, the ability to recall number facts instantly is called "fact power." Instant recall of the addition and subtraction facts will become a powerful tool in computation with multidigit numbers, such as 29 + 92.

Math Tools

Your child will be using **Fact Triangles** to practice and review addition and subtraction facts. Fact Triangles are a "new and improved" version of flash cards; the addition and subtraction facts shown are made from the same three numbers, helping your child understand the relationships among those facts. The Family Note on Home Link 2-7, which you will receive later, provides a more detailed description of Fact Triangles.

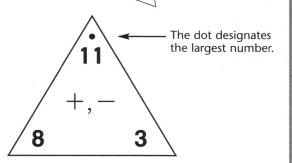

The dot designates the largest number.

A Fact Triangle showing the fact family for 3, 8, and 11

Vocabulary

Important terms in Unit 2:

label A unit, descriptive word, or phrase used to put a number or numbers in context. Using a label reinforces the idea that numbers always refer to something.

unit box A box that contains the label or unit of measure for the numbers in a problem. For example, in number stories involving children in the class, the unit box would be as follows:

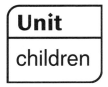

Unit
children

A unit box allows children to remember that numbers have a context without having to repeat the label in each problem.

number story A story involving numbers made up by children, teachers, or parents. Problems from the story can be solved with one or more of the four basic arithmetic operations.

number model A number sentence that shows how the parts of a number story are related. For example, 5 + 8 = 13 models the number story: "5 children skating. 8 children playing ball. How many children in all?"

fact power The ability to instantly recall basic arithmetic facts.

doubles fact The sum or product of the same two 1-digit numbers, such as 2 + 2 = 4 or 3 × 3 = 9.

turn-around facts A pair of addition (or multiplication) facts in which the order of the addends (or factors) is reversed, such as 3 + 5 = 8 and 5 + 3 = 8 (or 3 × 4 = 12 and 4 × 3 = 12). If you know an addition or multiplication fact, you also know its turn-around fact.

fact family A collection of four addition and subtraction facts, or multiplication and division facts, relating three numbers. For example, the addition/subtraction fact family for the numbers 2, 4, and 6 consists of:

$2 + 4 = 6$ $4 + 2 = 6$
$6 - 4 = 2$ $6 - 2 = 4$

The multiplication/division fact family for the numbers 2, 4, and 8 consists of:

$2 \times 4 = 8$ $8 \div 2 = 4$
$4 \times 2 = 8$ $8 \div 4 = 2$

Frames-and-Arrows diagram A diagram used to represent a number sequence, or a list of numbers ordered according to a rule. A Frames-and-Arrows diagram has frames connected by arrows to show the path from one frame to the next. Each frame contains a number in the sequence; each arrow represents a rule that determines which number goes in the next frame.

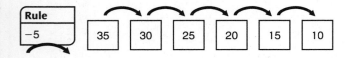

"What's My Rule?" problem A problem in which number pairs are related to each other according to a rule or rules. A rule can be represented by a **function machine.**

in	out
3	8
5	10
8	13

"What's My Rule?" table

Function machine In *Everyday Mathematics,* an imaginary device that receives input numbers and pairs them with output numbers according to a set rule.

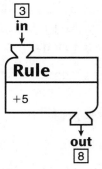

Do-Anytime Activities

To work with your child on the concepts taught in this unit and in previous units, try these interesting and rewarding activities:

1. Talk with your child about why it is important to learn basic facts.

2. Create addition and subtraction stories about given subjects.

3. Have your child explain how to use a facts table.

4. As you discover which facts your child is having difficulty mastering, make a Fact Triangle using the three numbers of that fact family.

5. Name a number and ask your child to think of several different ways to represent that number. For example, 10 can be represented as $1 + 9$, $6 + 4$, $12 - 2$, and so on.

10	
ten	$12 - 2$
$1 + 9$	$6 + 4$
diez	$10 - 0$

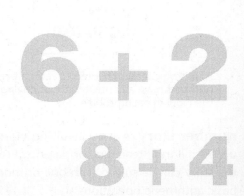

Building Skills through Games

In Unit 2, your child will practice addition facts and find equivalent names for numbers by playing the following games.

Beat the Calculator

A "Calculator" (a player who uses a calculator to solve the problem) and a "Brain" (a player who solves the problem without a calculator) race to see who will be first to solve addition problems.

Domino Top-It

Each player turns over a domino and finds the total number of dots. The player with the larger total then takes both dominoes from that round.

Doubles or Nothing

Each player adds numbers across each row, down each column, and along each diagonal in a grid for each round, circles identical sums, and finds the total of doubles as a score for the round.

Name That Number

Each player turns over a card to find a number that must be renamed using any combination of five faceup cards.

4	10	8	12	2		6
4	10	8	12	2		6

$$6 = 8 - 2$$
$$6 = 10 - 4$$
$$6 = 4 + 2$$

As You Help Your Child with Homework

As your child brings home assignments, you may want to go over the instructions together, clarifying them as necessary. The answers listed below will guide you through this unit's Home Links.

Home Link 2·1

2. 8 **3.** 18 **4.** 7 **5.** 16

Home Link 2·2

Home Link 2·3

1. a. 4 **b.** 10 **c.** 0 **d.** 14 **e.** 6
f. 16 **g.** 12 **h.** 18 **i.** 2 **j.** 8

3. a. 9 **b.** 9 **c.** 17 **d.** 13 **e.** 5
f. 15 **g.** 11

Home Link 2·4

1. a. 7 **b.** 11 **c.** 7 **d.** 7 **e.** 11 **f.** 7

2. a. 8 **b.** 5 **c.** 6 **d.** 3 **e.** 7 **f.** 9

3. a. 11 **b.** 15 **c.** 16 **d.** 10 **e.** 14 **f.** 15
g. 17 **h.** 14 **i.** 18 **j.** 16 **k.** 13 **l.** 17

Home Link 2·5

Home Link 2·6

2. $9 + 6 = 15$; $6 + 9 = 15$; $15 - 6 = 9$; $15 - 9 = 6$

3. $8 + 7 = 15$; $7 + 8 = 15$; $15 - 7 = 8$; $15 - 8 = 7$

4. $5 + 9 = 14$; $9 + 5 = 14$; $14 - 9 = 5$; $14 - 5 = 9$

5. 13 **6.** 14 **7.** 12 **8.** 16

Home Link 2·10

1. Rule +6
3 9 15 21 27

2. Rule −3
31 28 25 22 19

3. Rule +5
7 12 17 22 27

4. Rule +3
8 11 14 17 20

5. Rule −2
40 38 36 34 32

6. Rule +10
20 30 40 50 60

Home Link 2·11

1. Rule +9

in	out
1	10
4	13
6	15
8	17
5	14

2. Rule −8

in	out
10	2
12	4
9	1
14	6
8	0

3. Rule +6

in	out
4	10
6	12
3	9
9	15
0	6

4. Rule +5

in	out
8	13
4	9
13	18
5	10
Answers vary.	

5. 18; 5

HOME LINK 2·1

Addition Number Stories

Family Note

Before beginning this Home Link, review the vocabulary from the Unit 2 Family Letter with your child: **number story, label, unit box,** and **number model.** Encourage your child to make up and solve number stories and to write number models for the stories. Stress that the answer to the question makes more sense if it has a label.

Please return this Home Link to school tomorrow.

MRB
108

1. Tell someone at home what you know about number stories, labels, unit boxes, and number models. Write an addition number story for the picture. Write the answer and a number model.

Unit
lions

Story: _____

_____

_____

_____

_____

Answer the question: _____
(unit)

Number Model: __ + __ = ____

Practice

2. 6 + 2 = ____

3. 11 + 7 = ____

4. 4
    + 3
    ___

5. 10
    + 6
    ___

HOME LINK 2·2 | Addition Facts

Family Note In class today, we continued working with addition stories. We reviewed shortcuts when adding 0 or 1 to a number. We also stressed the importance of memorizing the sum of two 1-digit numbers. Then we reinforced addition facts by playing a game called *Beat the Calculator*.

Please return this Home Link to school tomorrow.

Solve these addition fact problems.

2 + 4	0 + 0	5 + 4	1 + 4	2 + 5	3 + 2	1 + 9	3 + 6	4 + 4	1 + 1
2 + 0	3 + 5	5 + 1	1 + 4	9 + 2	0 + 7	2 + 3	2 + 2	7 + 2	3 + 4	2 + 8
6 + 2	1 + 6	5 + 5	0 + 6	4 + 3	0 + 5	1 + 8	4 + 6	5 + 3	4 + 0	3 + 1
0 + 8	6 + 6	8 + 2	9 + 0	3 + 3	7 + 1	2 + 6	1 + 3	5 + 2	6 + 1	0 + 4
2 − 1	2 + 9	6 + 2	6 + 4	0 + 1	4 + 2	6 + 3	0 + 2	5 + 1	1 + 2	2 + 7
4 − 5	7 + 0	6 + 2	9 + 3	1 + 5	0 + 9	1 + 7	1 + 5	7 + 3	0 + 6	6 + 5
9 − 1	8 + 0	6 + 2	8 + 3	1 + 0	6 + 0	3 + 3	0 + 3	3 + 8	3 + 7	

HOME LINK 2·7 **Fact Triangles**

Family Note

Fact Triangles are tools used to help build mental arithmetic skills. You might think of them as the *Everyday Mathematics* version of flash cards. Fact Triangles are more effective for helping children memorize facts, however, because of their emphasis on fact families. A **fact family** is a collection of related addition and subtraction facts that use the same 3 numbers. The fact family for the numbers 2, 4, and 6 consists of $2 + 4 = 6$, $4 + 2 = 6$, $6 - 4 = 2$, and $6 - 2 = 4$.

To use Fact Triangles to practice addition with your child, cover the number next to the large dot with your thumb.

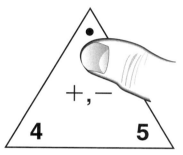

Your child tells you the addition fact: $4 + 5 = 9$ or $5 + 4 = 9$.

To use Fact Triangles to practice subtraction, cover one of the numbers in the lower corners with your thumb.

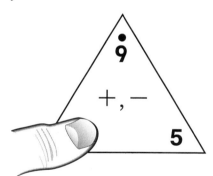 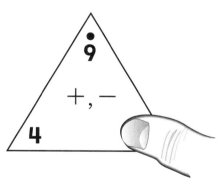

Your child tells you the subtraction facts: $9 - 5 = 4$ and $9 - 4 = 5$.

If your child misses a fact, flash the other two fact problems on the card and then return to the fact that was missed.

Example: Sue can't answer $9 - 5$. Flash $4 + 5$, then $9 - 4$, and finally $9 - 5$ a second time.

Make this activity brief and fun. Spend about 10 minutes each night over the next few weeks or until your child masters all of the facts. The work that you do at home will help your child develop an instant recall of facts and will complement the work that we are doing at school.

MRB 26 27

HOME LINK 2·8

Weighing Things

> **Family Note**
>
> Today we worked with a pan balance to compare the weights of objects. We used a spring scale to weigh objects up to 1 pound. We introduced the word *ounce* as a unit of weight for light objects.
>
> *Please return the **second page** of this Home Link to school tomorrow.*

1. Tell someone at home about how you used the pan balance to compare the weights of two objects.

2. Tell someone at home how you used the spring scale to weigh objects.

HOME LINK 2·8

Weighing Things *continued*

3. Look at the pairs of objects below. In each pair, circle the object that you think is heavier.

a.

Shoe

Marble

b.

Sock

Brick

c.

Feather

Tape Measure

4. Look at the objects below. Circle the objects that you think weigh less than 1 pound.

Pattern-Block Template

Scissors

Egg

Chair

Television

Pencil

Glasses

30

HOME LINK 2·13 | **Addition/Subtraction Facts** *continued*

Cut out the Fact Triangles. Show someone at home how you use them to practice adding and subtracting.

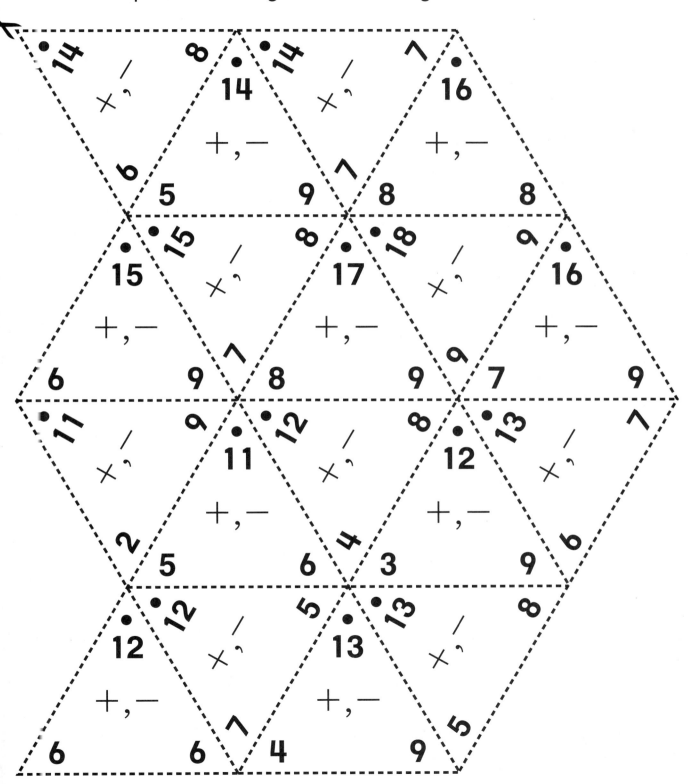

Unit 3: Family Letter

Place Value, Money, and Time

In Unit 3, children will read, write, and compare numbers from 0 through 999, working on concepts and skills built upon since *Kindergarten Everyday Mathematics.* Your child will review *place value,* or the meaning of each digit in a number. For example, in the number 52, the 5 represents 5 tens, and the 2 represents 2 ones.

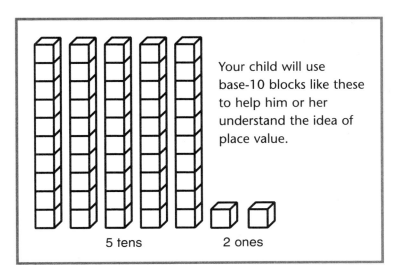

5 tens   2 ones

Your child will use base-10 blocks like these to help him or her understand the idea of place value.

Your child will also review money concepts, including finding the values of coins, identifying different coin combinations for the same amount, and making change.

43¢

43¢

43¢

Your child will read and record time using the hour and minute hands on an analog clock.

43

Vocabulary

Important terms in Unit 3:

analog clock A clock that shows time by the position of the hour and minute hands.

analog clock

digital clock A clock that shows time with numbers of hours and minutes, usually separated by a colon.

digital clock

data A collection of information, usually in the form of numbers. For example, the following data show the ages (in years) of six second graders: 6, 7, 6, 6, 7, 6.

middle number (median) The number in the middle of a list of data ordered from least to greatest or vice versa. For example, 5 is the middle number in the following ordered list:

2   3   ⑤   8   10

two-rule Frames and Arrows A Frames-and-Arrows diagram with two rules instead of just one, such as the following example.

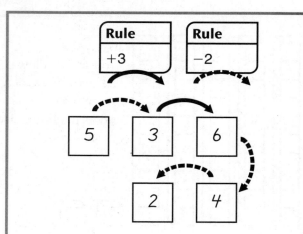

To go from the first square to the second square, use the rule for the dashed arrow.

$$5 - \mathbf{2} = 3$$

To go from the second square to the third square, use the rule for the solid arrow.

$$3 + \mathbf{3} = 6$$

Do-Anytime Activities

To work with your child on the concepts taught in this unit and in previous units, try these interesting and rewarding activities:

1. Have your child tell the time shown on an analog clock.

2. Draw an analog clock face without hands. Say a time and have your child show it on the clock face.

3. At the grocery store, give your child an item that costs less than $1.00. Allow your child to pay for the item separately. Ask him or her to determine how much change is due and to check that the change received is correct.

4. Gather a handful of coins with a value less than $2.00. Have your child calculate the total value.

5. Reinforce place value in 2- and 3-digit numbers. For example, in the number 694, the digit 6 means 6 hundreds, or 600; the digit 9 means 9 tens, or 90; and the digit 4 means 4 ones, or 4.

As You Help Your Child with Homework

As your child brings home assignments, you may want to go over the instructions together, clarifying them as necessary. The answers listed below will guide you through this unit's Home Links.

Home Link 3·1

1. **a.** 374   **b.** 507   **2.** 740

3. 936   **4.** 8; 0; 6   **5.** 2; 3; 1

Home Link 3·3

2. 6:30   **3.** 2:15   **4.** 9:00   **5.** 1:30

6.    **7.**

8.    **9.**

10. 13   **11.** 16   **12.** 6   **13.** 8

Home Link 3·4

1. | Rule | *Sample answers:*
   | Add 12 |

In	Out	Out in a different way
I	II ..	I ::.... .....
II	III	II ::::: :....
II	III	II ::::: ::..

2. | Rule |
   | Add 16 |

In	Out	Out in a different way
IIII .	IIIII	IIII ::::: ::....
.....	I :....	II .

Home Link 3·6

1. 40¢; 50¢; 55¢   2. 50¢; 45¢; 55¢

Home Link 3·7

5. 12   6. 14

7. 13   8. 10

Home Link 3·8

5¢; 35¢; 16¢; 5¢; 2¢; 52¢

1. 3.   2. 8   3. 7   4. 13

Building Skills through Games

In this unit, your child will practice addition and money skills by playing the following games:

Digit Game

Players turn over two cards and call out the largest number that can be made using those cards. The player with the higher number takes all the cards from that round.

Spinning for Money

Players "spin the wheel" to find out which coins they will take from the bank. The first player to exchange his or her coins for a dollar bill wins!

Dollar Rummy

Instead of three-of-a-kind, players of *Dollar Rummy* look for two cards that will add up to $1.00.

HOME LINK 3·4 | **"What's My Rule?" with Blocks**

Family Note Your child will complete the tables on this page by drawing tens and ones for 2-digit numbers. More than one picture can be drawn for a number. For example, to show 26, your child might draw 2 tens and 6 ones, 1 ten and 16 ones, or 26 ones. The symbol | stands for 10, and the symbol ▪ stands for 1.

Please return this Home Link to school tomorrow.

1. Draw simple pictures of base-10 blocks to complete the table.

Rule
Add 12

In	Out	Out in a Different Way
▪ ▪ ▪	| ▪ ▪ ▪ ▪ ▪	▪▪ ▪▪ ▪▪ ▪▪ ▪ ▪ ▪ ▪ ▪ ▪
|	|| ▪ ▪	
|| ▪ ▪ ▪ ▪		
|| ▪ ▪ ▪ ▪ ▪ ▪ ▪		

2. Write the rule. Then complete the table.

Rule

In	Out	Out in a Different Way
||||| ▪ ▪ ▪	||||| ▪ ▪ ▪ ▪ ▪ ▪ ▪ ▪ ▪ ▪	||||| ▪▪ ▪▪ ▪▪ ▪▪ ▪▪ ▪
| ▪ ▪ ▪ ▪ ▪ ▪ ▪ ▪	|| ▪▪ ▪▪ ▪ ▪ ▪ ▪ ▪ ▪	| ▪▪ ▪▪ ▪▪ ▪▪ ▪ ▪ ▪ ▪ ▪
|||| ▪		
▪ ▪ ▪ ▪ ▪		

53

HOME LINK 3·5

Pockets Bar Graph

Family Note   Help your child fill in the table below. Then display the data by making a **bar graph.**

Please return this Home Link to school tomorrow.

MRB 44

1. Pick five people. Count the number of pockets that each person's clothing has. Complete the table.

2. Draw a bar graph for your data. First, write the name of each person on a line at the bottom of the graph. Then color the bar above each name to show how many pockets that person has.

Name	Number of Pockets

How Many Pockets?

Number of Pockets

8+
7
6
5
4
3
2
1
0

Names

HOME LINK 3·7

Change at a Garage Sale

Family Note Encourage your child to make change by counting up. Using real coins and dollar bills will make this activity easier. *For example:*

◆ Start with the cost of an item—65 cents.

◆ Count up to the money given—$1.00.

One way to make change: Put down a nickel and say "70." Then put down 3 dimes and say "80, 90, 1 dollar." *Another way:* Put down 3 dimes and say "75, 85, 95." Then put down 5 pennies and say "96, 97, 98, 99, 1 dollar."

The Practice section in the Home Link provides a review of previously learned skills.

Please return this Home Link to school tomorrow.

Pretend you are having a garage sale. Do the following:

◆ Find small items in your home to "sell."

◆ Give each item a price less than $1.00.
Give each item a different price.

◆ Pretend that customers pay for each item with a $1 bill.

◆ Show someone at home how you would make change by counting up. Use Ⓟ, Ⓝ, Ⓓ, and Ⓠ.

◆ Show another way you can make change for the same item.

Example:

The customer buys __*a pen*__ for __65¢__ .

One way I can make change: _____ Ⓝ Ⓓ Ⓓ Ⓓ _____

Another way I can make change: __ Ⓓ Ⓓ Ⓓ Ⓟ Ⓟ Ⓟ Ⓟ Ⓟ __

HOME LINK 3·7 **Change at a Garage Sale** *continued*

1. The customer buys _____ for _____.

   One way I can make change: _____

   Another way I can make change: _____

2. The customer buys _____ for _____.

   One way I can make change: _____

   Another way I can make change: _____

3. The customer buys _____ for _____.

   One way I can make change: _____

   Another way I can make change: _____

4. The customer buys _____ for _____.

   One way I can make change: _____

   Another way I can make change: _____

Practice

5. $7 + 5 =$ ___

6. $8 + 6 =$ ___

7. $\begin{array}{r} 4 \\ + 9 \\ \hline \end{array}$

8. $\begin{array}{r} 3 \\ + 7 \\ \hline \end{array}$

HOME LINK 3·8 **Counting Up to Make Change**

> **Family Note** Help your child identify the amount of change that he or she would receive by "counting up" from the price of the item to the amount of money that was used to pay for the item. It may be helpful to act out the problems with your child using real coins and bills.
>
> *Please return this Home Link to school tomorrow.*

Complete the table.

I buy:	It costs:	I pay with:	My change is:
a bag of potato chips	70¢	Q Q Q	_____ ¢
a box of crayons	65¢	$1	_____ ¢
a pen	59¢	Q Q Q	_____ ¢
an apple	45¢	D D D D D	_____ ¢
a notebook	73¢	Q Q D D N	_____ ¢
a ruler	48¢	$1	_____ ¢
_____	_____	_____	_____ ¢
_____	_____	_____	_____ ¢

Practice

1. $12 - 9 =$ ____

2. $15 - 7 =$ ____

3. $\begin{array}{r} 13 \\ -6 \\ \hline \end{array}$

4. $\begin{array}{r} 17 \\ -4 \\ \hline \end{array}$

Unit 4: Family Letter

Addition and Subtraction

In Unit 4, children will use addition and subtraction stories to develop mental-arithmetic skills. Mental arithmetic is computation done in one's head or by drawing pictures, making tallies, or using manipulatives (counters, money, number lines, and number grids—no calculators, though). Children can also use their own solution strategies.

A second grader uses a number grid to solve 5 + 9.

1	2	3	4	⑤	6	7	8	9	10
11	12	13	⑭→⑮		16	17	18	19	20
21	22	23	24	25	26	27	28	29	30

Addition has two basic meanings: *putting together* and *changing to more.* In this unit, children will use **parts-and-total diagrams** and **change diagrams** to help them organize information in addition stories that either "put together" or "change to more." See the vocabulary section on page 87 to learn more about these diagrams.

I started at 5 and jumped ahead 10 to 15. But the problem said to add only 9, so I moved back 1 to 14.

Parts-and-Total Diagram

Total	
?	
Part	**Part**
20	16

Change Diagram

Change

Start		End
20	+6	?

Children will also develop estimation skills by solving problems that involve purchases. For example, your child will estimate whether $5.00 is enough to buy a pen that costs $1.69, a notebook that costs $2.25, and a ruler that costs 89¢.

In the last part of this unit, children will learn paper-and-pencil strategies for addition and will continue to gain hands-on experience with thermometers, money, tape measures, and rulers. Home Links 4-8 and 4-9, which you will receive later, will give you more information on the paper-and-pencil strategies that your child will be learning.

Please keep this Family Letter for reference as your child works through Unit 4.

Vocabulary

Important terms in Unit 4:

change-to-more number story A number story having a starting quantity that is increased so the ending quantity is more than the starting quantity.

For example: *Nick has 20 comic books. He buys 6 more. How many comic books does Nick have now?*

change diagram A device used to organize information in a change-to-more or change-to-less number story. The change diagram below organizes the information in Nick's comic book story above.

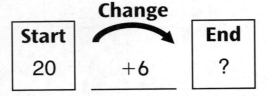

Start	Change	End
20	+6	?

mental arithmetic Computation done totally or partially in one's head, using a variety of strategies.

parts-and-total number story A number story in which two or more quantities (parts) are combined to form a total quantity. For example: *Carl baked 20 cookies. Sam baked 16 cookies. How many cookies did Carl and Sam bake in all?*

parts-and-total diagram A diagram used to organize information in a parts-and-total number story. The parts-and-total diagram below organizes the information in Carl's cookie story.

Total	
?	
Part	**Part**
20	16

estimate (1) An answer close to, or approximating, an exact answer. (2) To make an estimate.

algorithm A step-by-step set of instructions for doing something—for example, for solving addition or subtraction problems.

Building Skills through Games

In Unit 4, your child will practice addition and subtraction skills by playing the following games:

Addition Spin

A "Spinner" and a "Checker" take turns adding two numbers and checking the sum. After five turns, each player uses a calculator to find the sum of his or her scores. The player with the higher total wins.

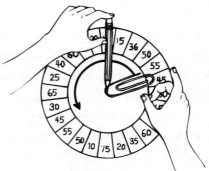

Name That Number

Each player turns over a card to find a number that must be renamed using any combination of five faceup cards.

Fact Extension Game

Players find sums of 2-digit numbers and multiples of ten.

Do-Anytime Activities

To work with your child on the concepts taught in this unit and in previous units, try these interesting and rewarding activities:

1. Encourage your child to show you addition and subtraction strategies as these concepts are developed during the unit.

2. Make up number stories involving estimation. For example, pretend that your child has $2.00 and that he or she wants to buy a pencil marked 64¢, a tablet marked 98¢, and an eraser marked 29¢. Help your child estimate the total cost of the three items (without tax) and determine whether there is enough money to buy them. If appropriate, you can also ask your child to estimate the amount of change due.

3. Look at weather reports in the newspaper and on television and discuss differences between high and low temperatures. Also note the differences between the Fahrenheit and Celsius scales.

As You Help Your Child with Homework

As your child brings home assignments, you may want to go over the instructions together, clarifying them as necessary. The answers listed below will guide you through this unit's Home Links.

Home Link 4·1

1. 18 grapes; 11 + 7 = 18

2. 38 cards; 30 + 8 = 38

3. 52 pounds; 42 + 10 = 52

4. 27   **5.** 80   **6.** 83

7. 10   **8.** 17   **9.** 70

10. 30   **11.** 66   **12.** 80

Home Link 4·2

1. 47 pounds; 17 + 30 = 47

2. 75 pounds; 45 + 30 = 75

3. 60 pounds; 15 + 45 = 60

4. 92 pounds; 17 + 45 + 30 = 92

Home Link 4·3

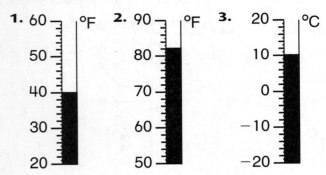

4. a. 14   **b.** 13   **c.** 6   **d.** 15

Home Link 4·4

1. 20°F   **2.** 34°F   **3.** 52°F

4. 96°F   **5.** 48°F   **6.** 73°F

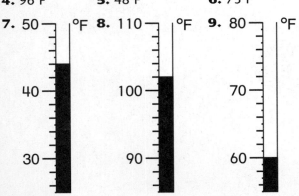

13. 70   **14.** 35   **15.** 97

16. 26   **17.** 50   **18.** 68

Home Link 4·5

1. no   **2.** yes   **3.** no   **4.** yes

5. 100   **6.** 46   **7.** 47

Home Link 4·6

1. 30 marbles; 20 + 10 = 30

2. 54 cookies; 30 + 24 = 54

3. 100   **4.** 140   **5.** 79   **6.** 83

7. 94   **8.** 77

Home Link 4·7

2. About 20 inches

Home Link 4·8

1. 76   **2.** 100   **3.** 83   **4.** 120

5. 98   **6.** 90   **7.** 93   **8.** 85

9. 71   **10.** 83   **11.** 169   **12.** 544

Home Link 4·9

1. 89   **2.** 108   **3.** 83   **4.** 94

5. 185   **6.** 363

HOME LINK 4·1

Change Number Stories

> **Family Note**
>
> Your child has learned about a device called a "change diagram" shown in the example below. Diagrams like this can help your child organize the information in a problem. When the information is organized, it is easier to decide which operation (+, −, ×, ÷) to use to solve the problem. Change diagrams are used to represent problems in which a starting quantity is increased or decreased. For the number stories on this Home Link, the starting quantity is always increased.
>
> *Please return the **second page** of this Home Link to school tomorrow.*
>
> MRB
> 116–118

Do the following for each number story on the next page:

◆ Write the numbers you know in the change diagram.

◆ Write "?" for the number you need to find.

◆ Answer the question.

◆ Write a number model.

Example: Twenty-five children are riding on a bus.
At the next stop, 5 more children get on.
How many children are on the bus now?

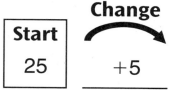

The starting number of children has been increased.

Answer: There are 30 children on the bus now.

Possible number model: 25 + 5 = 30

HOME LINK 4·1

Change Number Stories *continued*

1. Becky ate 11 grapes.
Later in the day she ate
7 more grapes.
How many grapes did she

eat in all? ____ grapes

Start	**Change**	End
	____	

Number model:

_____

2. Bob has 30 baseball cards.
He buys 8 more.
How many baseball
cards does Bob

have now? ____ cards

Start	**Change**	End
	____	

Number model:

_____

3. A large fish weighs
42 pounds.
A small fish weighs
10 pounds.
The large fish swallows the
small fish.
How much does the large

fish weigh now? ____ pounds

Start	**Change**	End
	____	

Number model:

_____

Practice

Add or subtract.

4. $20 + 7 =$ _____

5. _____ $= 40 + 40$

6. $3 + 80 =$ _____

7. $30 - 20 =$ _____

8. $47 - 30 =$ _____

9. $50 + 20 =$ _____

10. _____ $= 90 - 60$

11. $86 - 20 =$ _____

12. _____ $= 83 - 3$

HOME LINK 4·2 Parts-and-Total Number Stories

Family Note Today your child learned about another device to use when solving number stories. We call it a parts-and-total diagram. Parts-and-total diagrams are used to organize the information in problems in which two or more quantities (parts) are combined to form a total quantity.

*Please return the **second page** of this Home Link to school tomorrow.*

Large Suitcase
45 pounds

Small Suitcase
30 pounds

Backpack
17 pounds

Package
15 pounds

Use the weights shown in these pictures. Then do the following for each number story on the next page:

◆ Write the numbers you know in each parts-and-total diagram.

◆ Write "?" for the number you want to find.

◆ Answer the question.

◆ Write a number model.

Example: Twelve fourth graders and 23 third graders are on a bus. How many children in all are on the bus?

The parts are known. The total is to be found.

Total	
?	
Part	**Part**
12	23

Answer: 35 children

Possible number model: 12 + 23 = 35

69

Number Stories *continued*

1. You wear the backpack and carry the small suitcase. How many pounds do you

carry in all? ____ pounds

Total	
Part	**Part**

Number model:

_____

2. You carry the large suitcase and the small suitcase. How many pounds do you

carry in all? ____ pounds

Total	
Part	**Part**

Number model:

_____

3. You carry the package and the large suitcase. How many pounds do you

carry in all? ____ pounds

Total	
Part	**Part**

Number model:

_____

Try This

4. You wear the backpack and carry both of the suitcases. How many pounds do you

carry in all? ____ pounds

Total		
Part	**Part**	**Part**

Number model:

_____

HOME LINK
4·3

Reading a Thermometer

Family Note

In today's lesson, your child read temperatures on a real thermometer and on a thermometer pictured on a poster. The thermometers on this page show three different-size degree marks. The longest marks show 10-degree intervals, the medium-size marks show even-number degree intervals, and the shortest marks show odd-number degree intervals.

Help your child find the temperature shown by each thermometer by starting at a degree mark showing tens, counting the medium-size marks by 2s, and, if the temperature is at a short mark, counting 1 more.

Please return this Home Link to school tomorrow.

Circle the thermometer that shows the correct temperature.

1. 40°F

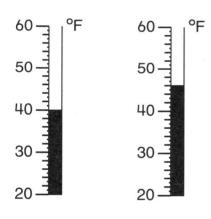

2. 82°F

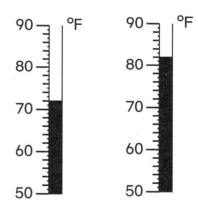

3. 10°C

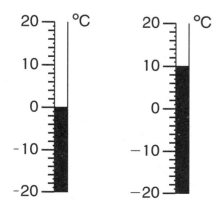

Practice

4. a. 6 + 8 = _____

  b. 7 + 6 = _____

  c. 9 + _____ = 15

  d. _____ = 8 + 7

Temperature

Family Note
In today's lesson, your child solved problems involving temperatures. On the thermometers on this Home Link, the longer degree marks are spaced at 2-degree intervals. Point to these degree marks while your child counts by 2s; 40, 42, 44, 46, 48, 50 degrees.

Problems 6 and 12 involve temperatures that are an odd number of degrees. Help your child use the shorter degree marks to get the correct answers.

Please return this Home Link to school tomorrow.

Write the temperature shown on each thermometer.

1.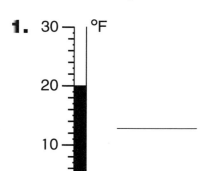

2.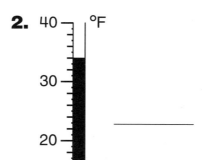

3.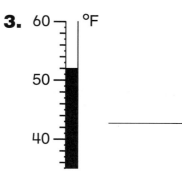

4.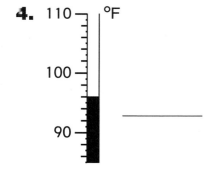

5.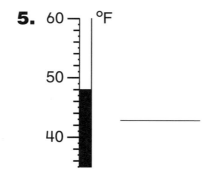

6.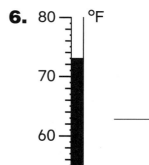

HOME LINK 4·4 **Temperature** *continued*

Fill in each thermometer to show the temperature.

7. Show 44°F.

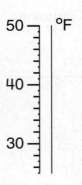

8. Show 102°F.

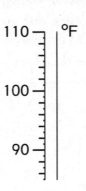

9. Show 60°F.

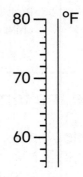

10. Show 56°F.

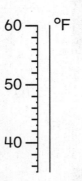

11. Show 38°F.

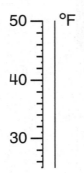

12. Show 27°F.

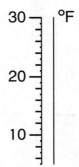

Practice

Add or subtract.

13. $30 + 40 =$ _____

14. $75 - 40 =$ _____

15. _____ $= 7 + 90$

16. _____ $= 46 - 20$

17. $\begin{array}{r} 53 \\ -3 \\ \hline \end{array}$

18. $\begin{array}{r} 60 \\ +8 \\ \hline \end{array}$

HOME LINK 4·5 | # Shopping at the Grocery Store

> **Family Note** Many problems in and out of the classroom require estimates rather than exact answers. In Problems 1–5 below, you need to know only whether the total cost is greater than $1.00 or less than $1.00; you do not need to know the exact total cost. In Problem 1, for example, help your child notice that the price of the can of frozen orange juice (98¢) is almost $1.00. Since a lemon is 10¢, your child could not buy both items.
>
> *Please return this Home Link to school tomorrow.*

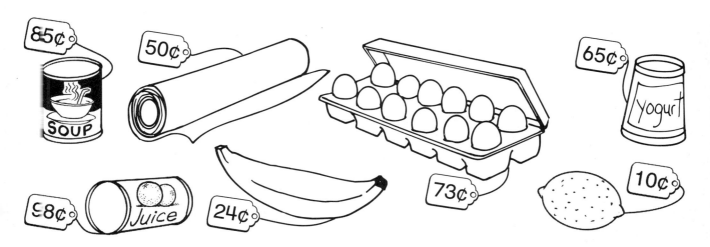

You have $1.00 to spend at the grocery store. Use estimation to answer each question.

| **Can you buy:** | **Circle *yes* or *no*.** |

1. a can of frozen orange juice and a lemon?   yes   no

2. a banana and a dozen eggs?   yes   no

3. a container of yogurt and a roll of paper towels?   yes   no

4. a lemon and a can of soup?   yes   no

Practice

Add or subtract.

5. 50 + 50 = _____   **6.** _____ = 6 + 40   **7.** _____ = 67 − 20

HOME LINK 4·6

Addition Number Stories

> **Family Note** In today's lesson, your child solved problems by adding 2-digit numbers mentally. For example, to find 34 + 23, you might first add the tens: 30 + 20 = 50. Then add the ones: 4 + 3 = 7. Finally, combine the tens and ones: 50 + 7 = 57.
>
> *Please return this Home Link to school tomorrow.*
>
> **MRB** 108 109 116 117

Try to solve Problems 1 and 2 mentally. Fill in the diagrams.
Then write the answers and number models.

1. Ruth had 20 marbles in her collection. Her brother gave her 10 more. How many marbles does Ruth have now?

Start	Change	End
	____	

Answer: _____
        (unit)

Number model:

_____

2. Tim baked 30 ginger snaps and 24 sugar cookies. How many cookies did he bake?

Total	
Part	**Part**

Answer: _____
        (unit)

Number model:

_____

Practice

Try to do each problem mentally. Then write the answer.

Unit

3. _____ = 40 + 60

4. 90 + 50 = _____

5. _____ = 70 + 9

6. 80 + 3 = _____

7. 30 + 64 = _____

8. _____ = 27 + 50

HOME LINK 4·7 | **Measuring to the Nearest Inch**

Family Note In today's lesson, your child measured the length, width, or height of objects to the nearest inch and centimeter. In later lessons, your child will make more precise measurements (such as measuring to the nearest half-inch).

Ask your child to show you how to measure the sections of the path on this page. Encourage your child to measure objects in your home.

Please return this Home Link to school tomorrow.

1. The ant will take this path to get to the picnic. Measure each part of the path to the nearest inch. If you do not have a ruler at home, cut out and use the ruler at the bottom of the page.

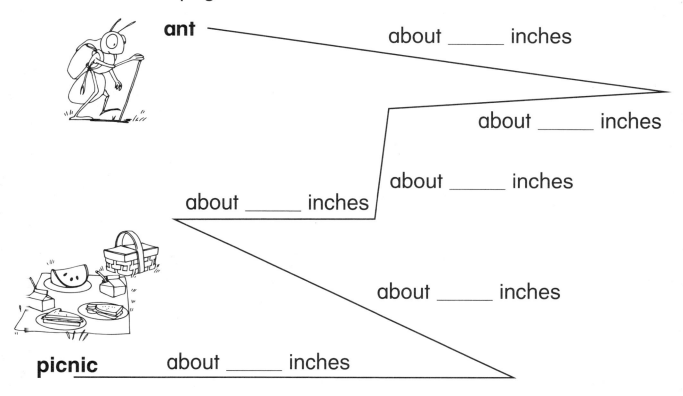

ant about _____ inches

about _____ inches

about _____ inches

about _____ inches

about _____ inches

picnic about _____ inches

2. What is the total length of the path? about _____ inches

```
|   |   |   |   |   |   |   |   |   |   |   |   |
0       1       2       3       4       5       6
inches
```

79

HOME LINK 4·8

Addition Strategies

Family Note

Everyday Mathematics encourages children to use a variety of strategies to solve computation problems. By doing so, children are developing a sense for numbers and operations rather than simply memorizing a series of steps.

We suggest that you give your child an opportunity to explore and choose addition strategies that he or she feels comfortable using. At some point, you may want to share the method that you know from your own school experience; please allow your child some time to use his or her own methods before doing so.

Below are three examples of methods that your child might use to solve 2-digit addition problems.

Counting On

$47 + 33 = ?$ ⟵———— "My problem"

$47\ 57\ 67\ 77$ ⟵———— "Start at 47. Count up 30 more."

$+\ 3$ ⟵———— "Add on 3 more."

80 ⟵———— "The answer is 80."

Combining Groups (1s, 10s, ...) Separately

$29 + 37 = ?$ ⟵———— "My problem"

$20 + 30 = 50$ ⟵———— "Add the tens."

$9 + 7 = \underline{16}$ ⟵———— "Add the ones."

66 ⟵———— "Put these together. The answer is 66."

Adjusting and Compensating

$52 + 29 = ?$ ⟵———— "My problem"

30 ⟵———— "30 is close to 29, just 1 more."

$52 + 30 = 82$ ⟵———— "52 plus 30 is 82."

$-\ 1$ ⟵———— "Take away 1, because I added 30 instead of 29."

81 ⟵———— "The answer is 81."

Encourage your child to use a ballpark estimate as a way to check whether an answer to a computation problem makes sense. For example, in $34 + 59$, 34 is close to 30 and 59 is close to 60. $30 + 60 = 90$ is your ballpark estimate. "90 is close to my answer 93, so 93 is a reasonable answer."

*Please return the **second page** of this Home Link to school tomorrow.*

81

Addition Strategies *continued*

Practice

Unit

Add.

1. 40 + 36 = _____   **2.** 20 + 80 = _____   **3.** _____ = 53 + 30

4. 60 + 60 = _____   **5.** _____ = 50 + 48   **6.** _____ = 70 + 20

Write a number model to show your ballpark estimate.

Add. Show your work in the workspaces.

Check your work.

7. Ballpark estimate: _____ 34 + 59	**8.** Ballpark estimate: _____ 17 + 68 =
9. Ballpark estimate: _____ 46 + 25 =	**10.** Ballpark estimate: _____ 56 + 27 =
11. Ballpark estimate: _____ 123 + 46 =	**12.** Ballpark estimate: _____ 318 + 226

HOME LINK 4·9

Place Value

Family Note

Your child is learning a method for addition that focuses on place value. The child is asked to first find a ballpark estimate. (For more on ballpark estimates see page 92 in the *My Reference Book*.)

Find 68 + 24

Ballpark estimate: 70 + 20 = 90

10s	1s
6	8
+ 2	4
8	0
+ 1	2
9	2

Add the tens (60 + 20 = 80) and write the sum.

Add the ones (8 + 4 = 12) and write the sum.

Combine the tens and ones (80 + 12 = 92) to find the final sum.

Encourage your child to use the correct place-value language when using this method. For example, when adding tens in the example, say "60 + 20 = 80," not "6 + 2 = 8." We only recently introduced this method, so allow plenty of time for practice before expecting your child to be able to use it easily.

Please return this Home Link to school tomorrow.

MRB 10

Write a number model for your ballpark estimate.
Find each sum.

Unit

1. Ballpark estimate:

_____

$$\begin{array}{r} 53 \\ + 36 \\ \hline \end{array}$$

2. Ballpark estimate:

_____

$$\begin{array}{r} 27 \\ + 81 \\ \hline \end{array}$$

3. Ballpark estimate:

_____

$$\begin{array}{r} 45 \\ + 38 \\ \hline \end{array}$$

Try This

4. Ballpark estimate:

_____

$$\begin{array}{r} 18 \\ + 76 \\ \hline \end{array}$$

5. Ballpark estimate:

_____

$$\begin{array}{r} 154 \\ + 31 \\ \hline \end{array}$$

6. Ballpark estimate:

_____

$$\begin{array}{r} 126 \\ + 237 \\ \hline \end{array}$$

HOME LINK 4·10

Unit 5: Family Letter

3-D and 2-D Shapes

Geometry is an important component of *Everyday Mathematics*. Studying geometry helps develop spatial sense and the ability to represent and describe the world. Instead of waiting until ninth or tenth grade, *Everyday Mathematics* introduces geometric fundamentals in Kindergarten and develops them over time. Children are thus prepared to study more advanced geometric topics later.

In Unit 5, children will consider five basic kinds of 3-dimensional shapes: prisms, pyramids, cylinders, cones, and spheres. To sort the shapes, children will explore similarities and differences among them. They will become familiar with both the names of shapes and the terms for parts of shapes.

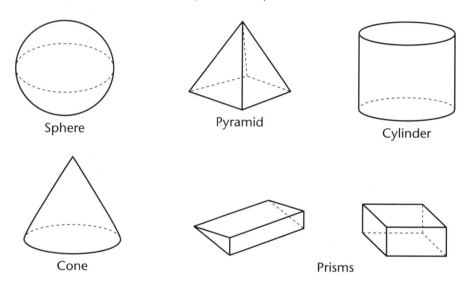

Sphere   Pyramid   Cylinder

Cone   Prisms

Children will also study **polygons,** or 2-dimensional shapes that form the flat surfaces of prisms and pyramids, as they look for examples in real life.

Later in the unit, children will explore **line symmetry** as they experiment with folding 2-dimensional shapes and matching the halves. Children will also cut out shapes and look for lines of symmetry in each shape. When children are given half of a shape, they will draw the missing half. Children will be asked to find symmetrical objects at home and in other places.

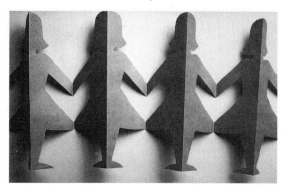

Please keep this Family Letter for reference as your child works through Unit 5.

Unit 5: Family Letter *cont.*

Vocabulary

The purpose of introducing children to the various shapes is to explore the characteristics of the shapes, not to teach vocabulary. This list is presented simply to acquaint you with some of the terms your child will be hearing in context in the classroom.

line segment A straight line joining two points. The two points are called endpoints of the segment.

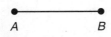

Line segment *AB* or *BA*

angle A figure formed by two rays or two line segments with a common endpoint called a vertex. The rays or segments are called the sides of the angle. The sides of a polygon form angles at each vertex.

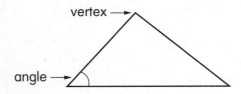

parallel lines Lines in plane that never meet. Two parallel lines are always the same distance apart.

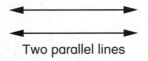

Two parallel lines

polygon A 2-dimensional figure formed by three or more line segments (*sides*) that meet only at their end points (*vertices*) to make a closed path. The sides may not cross one another.

polyhedron A 3-dimensional shape formed by *polygons* with their interiors (*faces*) and having no holes. Plural is *polyhedrons* or *polyhedra.* The following shapes are regular polyhedrons:

Tetrahedron

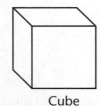
Cube

Octahedron

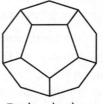

Dodecahedron   Icosahedron

face In *Everyday Mathematics* a flat surface on a 3-dimensional shape.

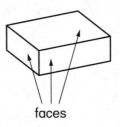

faces

vertex (corner) The point at which the ray of an angle, the sides of a polygon, or the edges of a polyhedron meet.

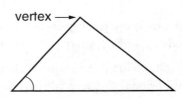

line symmetry A figure has line symmetry if a line can be drawn through it so that it is divided into two parts that are mirror images of each other. The two parts look alike but face in opposite directions.

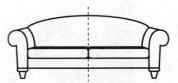

ray A part of a line starting at the ray's endpoint and continuing forever in one direction.

HOME LINK 4·10 | **Unit 5: Family Letter** *cont.*

Do-Anytime Activities

To work with your child on the concepts taught in this unit and in previous units, try these interesting and rewarding activities:

1. Together, look for 2-dimensional and 3-dimensional shapes in your home and neighborhood. Explore and name shapes and brainstorm about their characteristics. For example, compare a soup can and a tissue box. Talk about the differences between the shapes of the surfaces.

2. Use household items, such as toothpicks and marshmallows, straws and twist-ties, sticks, and paper to construct shapes like those shown below.

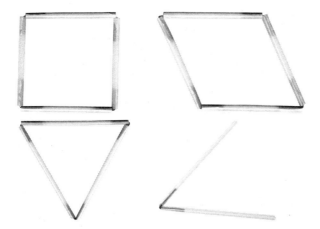

3. Look for geometric patterns in tile floors, quilts, buildings, and so on.

HOME LINK 4·10 | **Unit 5: Family Letter** *cont.*

As You Help Your Child with Homework

As your child brings home assignments, you may want to go over the instructions together, clarifying them as necessary. The answers listed below will guide you through this unit's Home Links.

Home Link 5·1

1.

2. The shapes all have 4 sides. **3.** Answers vary.

4. 66   **5.** 104   **6.** 58

Home Link 5·2

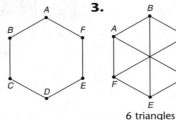

4. Answers vary.   **5.** 43   **6.** 44   **7.** 75

Home Link 5·3

1. 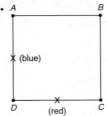   **2.** 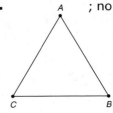 ; no

Home Link 5·5

1. the square   **2.** the rectangle

3. 9   **4.** 14   **5.** 3   **6.** 3

7. 20   **8.** 6   **9.** 97   **10.** 91

Home Link 5·6

1. 18   **2.** 27   **3.** 62   **4.** 96

Home Link 5·8

1.–3. Answers vary.   **4.** 12   **5.** 15

6. 16   **7.** 3   **8.** 5   **9.** 3

Building Skills through Games

In Unit 5, your child will practice addition and money skills by playing the following games:

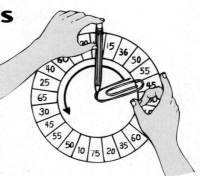

Addition Spin

Players "spin the wheel" twice and add the two selected numbers. Players check their partners' addition with a calculator.

Dollar Rummy

Instead of three-of-a-kind, players look for two cards that will add up to $1.00.

Beat the Calculator

A "Calculator" (a player who uses a calculator to solve a problem) and a "Brain" (a player who solves the problem without the calculator) race to see who will be the first to solve addition problems.

HOME LINK 5·1 "What's My Attribute Rule?"

Family Note Your child has been classifying shapes according to such rules as *only large shapes, only small red shapes,* or *only triangles.* Help your child determine which shapes in Problem 1 fit the rule by checking those shapes against the shapes below. What do all the shapes that fit the rule have in common? (They all have 4 sides.) Once your child thinks she or he knows the rule, check that rule against the shapes that do NOT fit the rule. Do any of those shapes follow the proposed rule?

Please return this Home Link to school tomorrow.

These shapes fit the rule.

  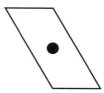

These shapes do NOT fit the rule.

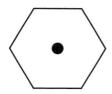   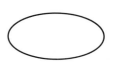

1. Which of these shapes fit the rule? Circle them.

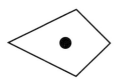 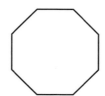  

2. What is the rule? _____

3. Draw a new shape that fits the rule.

Practice

4. 46 + 20 = _____   **5.** 74 + 30 = _____   **6.** 27 + 31 = _____

HOME LINK 5·2 Line Segments

Use a straightedge to draw line segments.

1. Draw these line segments:

$\overline{AC}$
$\overline{CE}$
$\overline{EA}$
$\overline{BF}$
$\overline{BD}$
$\overline{DF}$

A

B   F

C   D   E

2. Draw these line segments:

$\overline{AB}$
$\overline{BC}$
$\overline{CD}$
$\overline{DE}$
$\overline{EF}$
$\overline{FA}$

A

B   F

C   D   E

3. Draw the following line segments:

$\overline{AB}$, $\overline{BC}$
$\overline{CD}$, $\overline{DE}$
$\overline{EF}$, $\overline{FA}$
$\overline{AD}$, $\overline{FC}$
$\overline{BE}$

B

A   C

F   E   D

How many triangles are there? _____

4. Draw points on the back of this page. Label each point with a letter. Use a straightedge to connect the points with line segments to make polygons.

Practice

5. 23 + 20 = _____

6. 14 + 30 = _____

7. 45 + 30 = _____

91

HOME LINK 5·3 | Parallel Line Segments

Family Note Parallel line segments are always the same distance apart. They would never meet, even if they were extended forever in either or both directions. In Problem 1, line segment *DC* is parallel to line segment *AB*, and line segment *AD* is parallel to line segment *BC*. There are no parallel line segments in Problem 2.

*Please return the **top part** of this Home Link to school tomorrow.*

MRB
51

1. Draw line segments *AB*, *BC*, *CD*, and *DA*.

A •   • B

Put a red **X** on the line segment that is parallel to line segment *AB*.

Put a blue **X** on the line segment that is parallel to line segment *BC*.

D •   • C

2. Draw line segments *AB*, *BC*, and *CA*.

A
•

Is any line segment in your drawing parallel to line segment *AB*? _____

C •   • B

- -

Special Family Note In Lesson 5-6, your child will be studying 3-dimensional shapes. Help your child gather 3-dimensional objects for a class collection that we call the "Shapes Museum." You and your child might want to separate the objects you collect according to shape.

Shapes Museum

For the next few days, your class will collect things to put into a Shapes Museum. Starting tomorrow, bring items like boxes, soup cans, party hats, pyramids, and balls to school. Ask an adult for permission before bringing in these items. Make sure that the things you bring are clean.

HOME LINK 5·4 **Polygons**

Family Note In this lesson, your child has been learning the names of different polygons. A polygon is a closed figure made up of straight sides, and you can trace and come back to where you started without retracing or crossing any part. Different types of polygons are shown below. Examples of polygons can be found in real-life objects. For example, a stop sign is an octagon and this page is a rectangle. As your child cuts out pictures of polygons, discuss each shape. Count the sides and angles and try to name the polygons. Talk about how the polygons are alike and different.

Please return this Home Link to school tomorrow or as requested by the teacher.

1. Cut out pictures from newspapers and magazines that show triangles, quadrangles, and other polygons. Ask an adult for permission first.

2. Paste each picture on a sheet of paper.

3. Write the names of some of the polygons under the pictures.

4. Bring your pictures to school.

Triangles	**Quadrangles or Quadrilaterals**
Pentagons	**Hexagons**
Heptagons	**Octagons**

These are NOT polygons.

HOME LINK 5·5 | **Quadrangles**

Family Note In this lesson, your child has been learning about different types of quadrangles, or polygons that have 4 sides. Quadrangles are also called *quadrilaterals*. In Problems 1 and 2 below, three shapes have a common attribute that the fourth shape does not have. In Problem 1, the square is different, because it is the only quadrangle with 4 square corners. In Problem 2, the rectangle is different, because it is the only quadrangle that doesn't have 4 equal sides.

Please return this Home Link to school tomorrow.

MRB 55

1. Look at the number of square corners. Which quadrangle is different from the other three?

_____

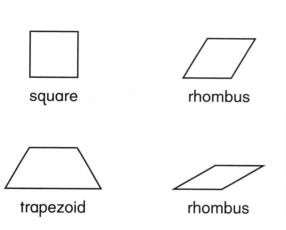

square   rhombus

trapezoid   rhombus

2. Look at the lengths of the sides. Which quadrangle is different from the other three?

_____

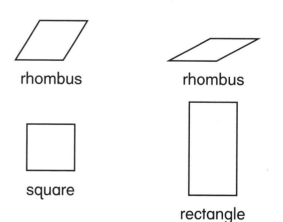

rhombus   rhombus

square

rectangle

Practice

3. 6 + 3 = _____

4. 5 + 9 = _____

5. 6 − 3 = _____

6. 8 − 5 = _____

7. 24 − 4 = _____

8. 56 − 50 = _____

9. 35 + 62 = _____

10. 25 + 66 = _____

HOME LINK 5·6

3-D Shapes

Family Note In this lesson, children have identified and compared 3-dimensional shapes. Our class also has created a Shapes Museum using the objects that children brought to school. Read your child's list of shapes. Together, find shapes to complete the list.

Please return this Home Link to school tomorrow.

MRB 56 57

On your way home, look for things that have these five shapes.

Make a list of things you see. Show your list to someone at home. Can you find any more shapes in your home? Add them to your list.

Prisms

_____

_____

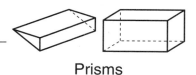

Prisms

Cones

_____

_____

Cone

Pyramids

_____

_____

_____

Pyramids

Spheres

_____

_____

_____

Sphere

Cylinders

_____

_____

_____

Cylinder

| **Practice** |

1. 10 + 8 = _____

2. 20 + 7 = _____

3. 42 + 20 = _____

4. 66 + 30 = _____

HOME LINK
5·7

Make a Triangular Pyramid

Family Note   Your child has used straws and twist-ties to construct pyramids with different-shape bases. The *base* can be a triangle, a rectangle, a pentagon, or another shape. Help your child construct a triangular pyramid (a pyramid with a triangle as the base) by using the cutout pattern below. After constructing the pyramid, ask your child the following questions:

◆ What is the shape of the base? (*A triangle*)

◆ How many edges does the pyramid have? (*6*)

◆ How many faces does the pyramid have? (*4*)

◆ How many vertices does the pyramid have? (*4*)

Please return this Home Link to school tomorrow.

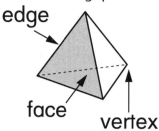

Ask someone at home to help you make a triangular pyramid out of this pattern.

1. Cut on the dashed lines.

2. Fold on the dotted lines.

3. Tape or glue tabs "inside" or "outside."

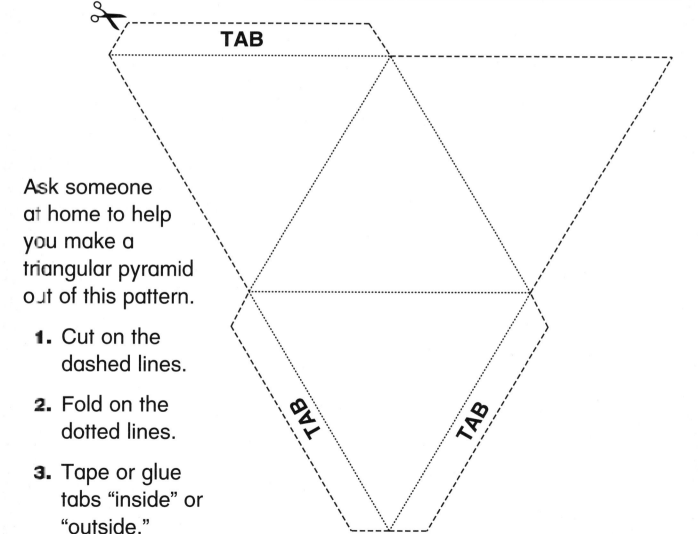

TAB

TAB   TAB

HOME LINK 5·8 **Symmetry Hunt**

Family Note In this lesson, your child has been determining whether shapes are symmetrical. A shape has *symmetry* if it has two halves that look alike but face in opposite directions. A *line of symmetry* divides the shape into two matching parts. Lines of symmetry are shown in the objects below. Help your child find other objects that are symmetrical. Remember that some shapes, such as the mirror below, may be symmetrical in more than one way.

Please return this Home Link to school tomorrow.

MRB 60

1. Ask someone to help you make a list of things at home that have symmetry. For example, you might list a window, a sofa, or a mirror.

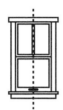
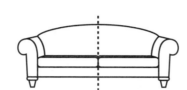
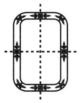

My list: _____

_____

2. Draw a picture of one thing on your list. Draw as many lines of symmetry as you can.

3. If you find pictures in books or magazines that show symmetry, bring them to school.

Practice

4. $4 + 8 =$ _____

5. $6 + 9 =$ _____

6. $8 + 8 =$ _____

7. $8 - 5 =$ _____

8. $9 - 4 =$ _____

9. $7 - 4 =$ _____

HOME LINK 5·9 ## Unit 6: Family Letter

Whole-Number Operations and Number Stories

In Unit 6, children will take another look at the addition and subtraction diagrams that were introduced in Unit 4.

Later in this unit, children will strengthen their understanding of multiplication and division as they act out number stories using manipulatives and arrays, complete diagrams to show the relationships in multiplication problems, and then begin to record corresponding number models.

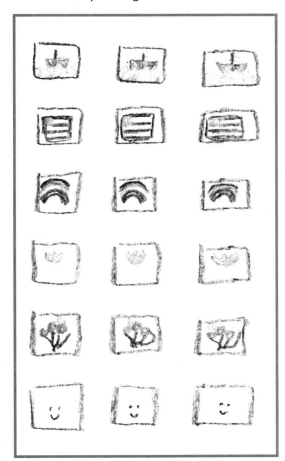

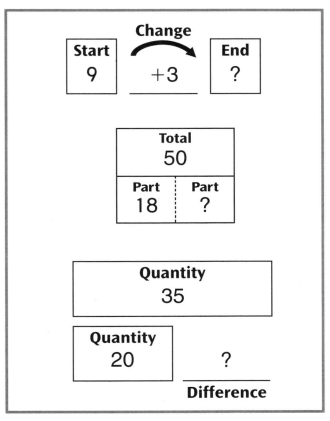

above: addition and subtraction diagrams

left: A child uses an array to solve the following problem: A sheet of stamps has 6 rows. Each row has 3 stamps. How many stamps are on a sheet?

below: multiplication diagram

boxes	marbles per box	marbles in all
3	7	?

Please keep this Family Letter for reference as your child works through Unit 6.

Vocabulary

Important terms in Unit 6:

comparison number story A number story that involves the difference between two quantities. For example: Ross sold 12 cookies. Anthony sold 5 cookies. How many more cookies did Ross sell?

comparison diagram A diagram used to organize the information from a comparison number story. For example, the diagram below organizes the information from Anthony's cookie story above.

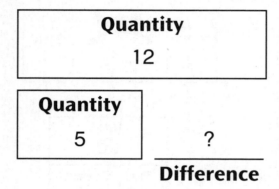

rectangular array An arrangement of objects into rows and columns. For example, 20 pencils could be arranged in 4 rows of 5 pencils each.

multiples of a number The product of the number and a counting number. For example, multiples of 2 are 2, 4, 6, 8, and 10 because $2 \times 1 = 2$, $2 \times 2 = 4$, $2 \times 3 = 6$, and so on.

remainder The amount left over when one number is divided by another number. For example, if 20 pencils are shared equally by 6 people, each person gets 3 pencils, and 2 are left over. The remainder is 2.

Do-Anytime Activities

To work with your child on the concepts taught in this unit and in previous units, try these interesting and rewarding activities:

1. Have your child show you how making an array or making equal groups can help solve multiplication number stories. Use common objects, such as buttons or pennies, to act out the stories.

2. Also try the opposite: Draw or make arrays and multiples of equal groups. Have your child make up and solve number stories to go with them.

3. Discuss equal-sharing (division) stories. For example, use objects (such as pennies) to portray a situation like the following: We have 7 cookies to divide equally among 3 people. How many whole cookies will each person get? (2) How many cookies will be left over? (1)

Building Skills through Games

In Unit 6, your child will practice addition, subtraction, and multiplication skills by playing the following games:

Three Addends

Players draw three cards, write addition models of the numbers they've picked, and solve the problems.

Addition Top-It

Each player turns over two cards and calls out their sum. The player with the higher sum then takes all the cards from that round.

Array Bingo

Players roll the dice and find an *Array Bingo* card with the same number of dots. Players then turn that card over. The first player to have a row, column, or diagonal of facedown cards calls out "Bingo!" and wins the game.

Number-Grid Difference Game

Players subtract 2-digit numbers using the number grid.

Fact Extension Game

Players find sums of 2-digit numbers and multiples of ten.

As You Help Your Child with Homework

As your child brings home assignments, you may want to go over the instructions together, clarifying them as necessary. The answers listed below will guide you through this unit's Home Links.

Home Link 6·1
Sample answers:
  1. $13 + 6 + 7 = 26$
  2. $22 + 8 + 5 = 35$
  3. $15 + 9 + 25 = 49$
  4. $29 + 11 + 6 = 46$
  5. 69   **6.** 70   **7.** 62
  8. 83   **9.** 148   **10.** 190

Home Link 6·2
  1. $19; 29 − 10 = 19
  2. 15 fewer laps; $20 + 15 = 35$
  3. June 22; $10 + 12 = 22$

  4. $\begin{array}{r} 90 \\ +11 \\ \hline 101 \end{array}$ **5.** $\begin{array}{r} 40 \\ +15 \\ \hline 55 \end{array}$ **6.** $\begin{array}{r} 80 \\ +7 \\ \hline 87 \end{array}$

Home Link 6·3
  1.

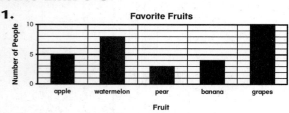

Favorite Fruits

  2. grapes; pear

Home Link 6·4
  1. 30
  2. 28
  3. 20

Home Link 6·5
  1. 58; 41 cubes left; $58 − 17 = 41$
  2. 26; 8 cubes left; $26 − 18 = 8$
  3. 43; 18 cubes left; $43 − 25 = 18$
  4. 39; 7 cubes left; $39 − 32 = 7$
  5. 61; 14 cubes left; $61 − 47 = 14$

Home Link 6·6
  1. 4 rows; 5 Xs in each row; 20

Home Link 6·7
  1. 3; 18   **2.** 2; 8   **3.** 10; 80

Home Link 6·8
  1. 24
  2. 35

Home Link 6·9
  1. Total = 21; $7 \times 3 = 21$
  2. Total = 60; $6 \times 10 = 60$
  3. 5 rows; 6 dots in each row; 30
  4. 3 rows; 9 squares per row; 27
  5. 6 rows; 6 squares in each row; 36

Home Link 6·10
  3. by 2 people: 9¢ per person; 1¢ remaining
     by 3 people: 6¢ per person; 1¢ remaining
     by 4 people: 4¢ per person; 3¢ remaining

HOME LINK 6·1 | ## Adding Three Numbers

> **Family Note**   Sometimes the order in which you add numbers can make it easier to find the sum. For example, when adding 17, 19, and 23, some people may first calculate 17 + 23, which equals 40, and then add 19 *(40 + 19 = 59)*. For Problems 1–4, help your child look for easy combinations. Before working on Problems 5–10, you might go over the example with your child.
>
> *Please return this Home Link to school tomorrow.*

For each problem:

◆ Think about an easy way to add the numbers.

◆ Write a number model to show the order in which you are adding the numbers.

◆ Find each sum. Tell someone at home why you added the numbers in that order.

1.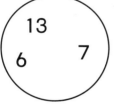

Number model:

_____ + _____ + _____ = _____

2.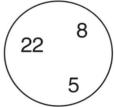

Number model:

_____ + _____ + _____ = _____

3.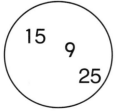

Number model:

_____ + _____ + _____ = _____

4.

Number model:

_____ + _____ + _____ = _____

HOME LINK 6·1   # Adding Three Numbers *continued*

Add. Use the partial-sums method.

Example:

$$\begin{array}{r} 33 \\ 42 \\ + \ 11 \end{array}$$

Add the tens.	→ *(30 + 40 + 10)*	→	80
Add the ones.	→ *(3 + 2 + 1)*	→	6
Add the partial sums.	→ *(80 + 6)*	→	86

Practice

5.
$$\begin{array}{r} 23 \\ 32 \\ + \ 14 \\ \hline \end{array}$$

6.
$$\begin{array}{r} 14 \\ 29 \\ + \ 27 \\ \hline \end{array}$$

7.
$$\begin{array}{r} 8 \\ 19 \\ + \ 35 \\ \hline \end{array}$$

8.
$$\begin{array}{r} 46 \\ 25 \\ + \ 12 \\ \hline \end{array}$$

9.
$$\begin{array}{r} 40 \\ 45 \\ + \ 63 \\ \hline \end{array}$$

10.
$$\begin{array}{r} 9 \\ 85 \\ + \ 96 \\ \hline \end{array}$$

HOME LINK 6·2

Comparison Number Stories

Family Note

Today your child learned about a device that is useful when solving number stories. We call it a comparison diagram. Diagrams like these can help your child organize the information in a problem. When the information is organized, it is easier to decide which operation (+, −, ×, or ÷) to use to solve the problem.

Comparison diagrams are used to represent problems in which two quantities are given and the question is how much more or less one quantity is than the other (the difference).

Example 1: There are 49 fourth graders and 38 third graders. How many more fourth graders are there than third graders?

Note that the number of fourth graders is being compared with the number of third graders.

- *Answer:* There are 11 more fourth graders than third graders.

- *Possible number models:* Children who think of the problem in terms of subtraction will write 49 − 38 = 11. Other children may think of the problem in terms of addition: "Which number added to 38 will give me 49?" They will write the number model as 38 + 11 = 49.

Quantity
49 fourth graders

Quantity	
38 *third graders*	?
	Difference

Your child may write words in the diagram as a reminder of what the numbers mean.

Example 2: There are 53 second graders. There are 10 more second graders than first graders. How many first graders are there?

Note that sometimes the difference is known and that one of the two quantities is unknown.

- *Answer:* There are 43 first graders.

- *Possible number models:*
  53 − 10 = 43 or 10 + 43 = 53

Quantity
53

Quantity	
?	10
	Difference

For Problems 1 and 2, ask your child to explain the number model that he or she wrote. Also ask your child to explain the steps needed to solve Problems 4–6.

*Please return the **second page** of this Home Link to school tomorrow.*

MRB
110 111

HOME LINK 6·2 **Comparison Stories** *continued*

In each number story:

◆ Write the numbers you know in the comparison diagram.

◆ Write ? for the number you want to find.

◆ Solve the problem. Then write a number model.

---

1. Ross has $29. Omeida has $10.

Ross has $_____ more than Omeida.

Number model: _____

Quantity

Quantity

Difference

---

2. Omar swam 35 laps in the pool. Anthony swam 20 laps.

Anthony swam _____ fewer laps than Omar.

Number model: _____

Quantity

Quantity

Difference

---

3. Claudia's birthday is June 10. Tisha's birthday is 12 days later.

Tisha's birthday is June _____.

Number model: _____

Quantity

Quantity

Difference

Practice

Add. Use the partial-sums method.

Unit

4.   39
   + 62

5.   48
   + 7

6.   33
   + 54

HOME LINK 6·3 | Graphing Data

Family Note The class has been collecting and graphing data about favorite foods. Ask your child about the graph he or she made in class. In the table below, help your child count the tally marks below the name of each fruit. To decide how high up to color each bar, your child could lay a straightedge across the columns.

Please return this Home Link to school tomorrow.

MRB 44 45

In a survey, people were asked to name their favorite fruit. The table below shows the results.

apple	watermelon	pear	banana	grapes
⊬ℍ	⊬ℍ ///	///	////	⊬ℍ ⊬ℍ

1. Make a bar graph that shows how many people chose each fruit. The first bar has been colored for you.

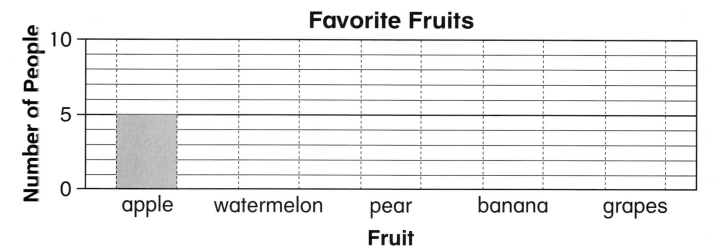

2. Which fruit is the most popular? _____

Which fruit is the least popular? _____

What is your favorite kind of fruit? _____

113

HOME LINK 6·4 | # Number Stories and Diagrams

> **Family Note** In today's lesson, your child used diagrams to solve number stories. Listen to your child's stories. Ask your child to explain how each story relates to both the diagram and the number model. (See Home Link 5.10: *Unit 6 Family Letter* for information about number stories and diagrams.)
>
> *Please return this Home Link to school tomorrow.*
>
> **MRB** 109–118

Write number stories to match each diagram. Then finish the number model. Tell your stories to someone at home.

1.

Unit
building blocks

Start	Change +6	End
24	+6	?

_____

_____

_____

Finish the number model: 24 + 6 = _____

2.

Unit
books

Total	
Part	**Part**

_____

_____

_____

Finish the number model: 15 + 13 = _____

HOME LINK 6·4

Number Stories and Diagrams *cont.*

3.

Unit
bananas

Quantity

Quantity

_____
Difference

_____

_____

Finish the number model: 28 − 8 = _____

4.

Unit
baseball cards

Total	
Part	**Part**

_____

_____

Write a number model for your story.

Number model: _____

HOME LINK 6·5 Subtracting with Base-10 Blocks

Family Note

In this lesson, your child found the answers to subtraction problems by using longs and cubes to represent tens and ones, respectively.

This will help your child understand the concept of subtraction before he or she learns to subtract using a step-by-step procedure, or algorithm, with paper and pencil. When you see the problems on this Home Link, you may be eager to teach your child to subtract the way you were taught. Please wait—the introduction of a formal algorithm for subtraction will be taught later in second grade.

Please return this Home Link to school tomorrow.

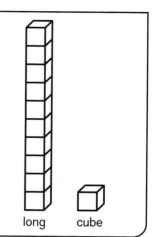

long   cube

Show subtraction by crossing out cubes.

Example:

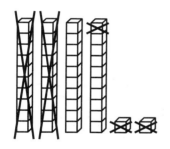

How many cubes are shown as separate cubes and as part of the longs? __42__

Cross out (subtract) 23 cubes. How many cubes are left? __19__

Number model:

__42__ – __23__ = __19__

1.

How many cubes are shown in all? _____

Cross out (subtract) 17 cubes. How many cubes are left? _____

Number model:

_____ – _____ = _____

117

HOME LINK 6·5

Subtracting with Blocks *continued*

2.

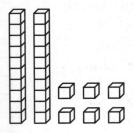

How many cubes
are shown in all? _____

Cross out (subtract)
18 cubes. How
many cubes are left? _____

Number model:

_____ − _____ = _____

3.

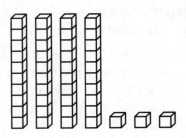

How many cubes
are shown in all? _____

Cross out (subtract)
25 cubes. How
many cubes are left? _____

Number model:

_____ − _____ = _____

4.

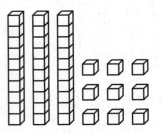

How many cubes
are shown in all? _____

Cross out (subtract)
32 cubes. How
many cubes are left? _____

Number model:

_____ − _____ = _____

5.

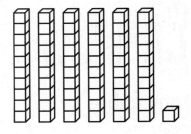

How many cubes
are shown in all? _____

Cross out (subtract)
47 cubes. How
many cubes are left? _____

Number model:

_____ − _____ = _____

HOME LINK 6·6 | How Many?

1. Show someone at home this array.

```
X X X X X
X X X X X
X X X X X
X X X X X
```

How many rows? _____

How many **X**s in each row? _____

How many **X**s in all? _____

2. Draw an array of 16 **X**s.

How many rows? _____

How many **X**s in each row? _____

3. Draw an array of 24 **X**s.

How many rows? _____

How many **X**s in each row? _____

4. Draw a different array of 24 **X**s.

How many rows? _____

How many **X**s in each row? _____

HOME LINK 6·7 | # How Many?

Family Note In today's lesson, your child learned that multiplication is an operation used to find the total number of things in several equal groups. As you help your child solve the following problems, emphasize that each group has the same number of things. Your child can use objects, draw pictures, count, or use any other helpful devices to find the answers.

Please return this Home Link to school tomorrow.

Example:

How many apples in 4 packages?

| ɫɫɫ   ɫɫɫ   ɫɫɫ   ɫɫɫ |
| 5 + 5 + 5 + 5 = 20 |

There are 20 apples in 4 packages.

---

1. △ △ △ △ △ △

How many sides on each triangle? _____ sides

How many sides in all? _____ sides

2.

How many wheels on each bike? _____ wheels

How many wheels in all? _____ wheels

---

3.    How many fingers for each person?

_____ fingers

How many fingers in all?

_____ fingers

HOME LINK 6·8 | Arrays

Family Note In this lesson, your child solved multiplication problems about arrays, which are rectangular arrangements of objects in rows and columns. Encourage your child to use counters, such as pennies or buttons, while working on the following exercises.

Please return this Home Link to school tomorrow.

Tell someone at home what you know about arrays.

1. Look at the array and fill in the blank.

● ● ● ● ● ●
● ● ● ● ● ●
● ● ● ● ● ●
● ● ● ● ● ●

4 rows of dots

6 dots in each row

_____ dots in all.

2. Draw an array of dots. Your array should have

5 rows of dots

7 dots in each row

That's _____ dots in all.

3. Draw an array of 12 dots.

Telephone:
a 4-by-3 array

Muffins:
a 3-by-2 array

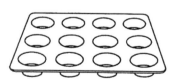

Muffins:
a 3-by-4 array

Tic-tac-toe Grid:
a 3-by-3 array

Checkerboard:
an 8-by-8 array

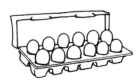

Eggs:
a 2-by-6 array

123

HOME LINK 6·9 Arrays

Family Note In this lesson, your child continued to work with arrays to develop multiplication concepts. Your child described each array by naming the number of rows, the number of items in each row, and the total number of items in the array. Your child wrote number models to describe arrays. In the example, an array with 2 rows of 4 dots can be described using the number model 2 × 4 = 8.

Please return this Home Link to school tomorrow.

Show an array for the numbers that are given. Find the total number of dots in the array. Complete the number model.

Example:

Numbers: 2, 4

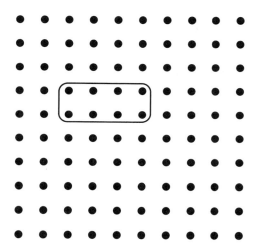

Total: __8__

Number model:

__2__ × __4__ = __8__

1. Numbers: 7, 3

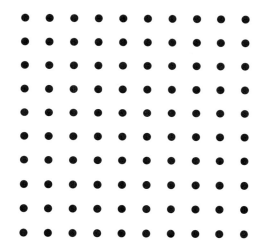

Total: _____

Number model:

_____ × _____ = _____

Arrays *continued*

2. Numbers: 6, 10

Total: _____

Number model:

_____ × _____ = _____

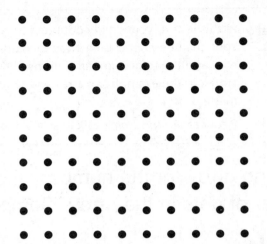

Answer the questions about each array.

3.

How many rows? _____

How many
dots in each row? _____

How many
dots in the array? _____

4.

How many rows? _____

How many
squares per row? _____

How many
squares in the array? _____

5.

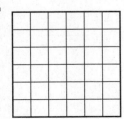

How many rows? _____

How many squares in each row? _____

How many squares in the array? _____

HOME LINK 6·10 Division

1. Have someone at home give you a group or handful of small items, such as raisins, buttons, or popcorn. Show how you can divide the items equally among your family members. Are any items left over?

Make a record of what you did. Be ready to tell about it in class.

I shared _____ (how many?) items equally among _____ people.

Each person got _____. There were _____ left over.

2. Do this again with some other kind of item.

I shared _____ items equally among _____ people.

Each person got _____. There were _____ left over.

3. 19 cents shared equally

by 2 people	by 3 people	by 4 people
_____¢ per person	_____¢ per person	_____¢ per person
_____¢ remaining	_____¢ remaining	_____¢ remaining

127

Unit 7: Family Letter

Patterns and Rules

In Unit 7, children will concentrate on number patterns, computational skills, and the application of mathematics through the use of data. They will continue to use the 100-grid to support their numeration skills. Children will also explore the patterns of doubling and halving numbers, which will help prepare them for multiplication and division.

Computational work will be extended to several 2-digit numbers and to the subtraction of 1- and 2-digit numbers from multiples of 10.

Children will learn to find complements of tens; that is, they will answer such questions as "What must I add to 4 to get to 10? What must I add to 47 to get to 50?" or "How many tens are needed to get from 320 to 400?"

Children will also collect and work with real-life data about animals, adults, and themselves. For example, they will collect data by measuring the lengths of their standing long jumps and then find the median jump length for the class.

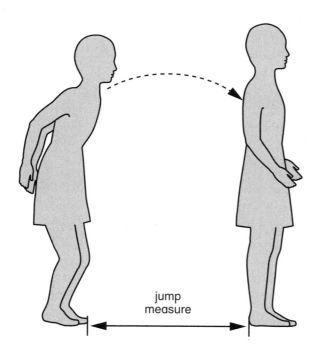

jump
measure

Please keep this Family Letter for reference as your child works through Unit 7.

Vocabulary

Important terms in Unit 7:

median (middle number) The number in the middle of a list of data ordered from least to greatest or vice versa. For example, 35 is the middle number in the following ordered list.

$$30, 32, 32, 35, 36, 38, 40$$

frequency The number of times an event or value occurs in a set of data. For example, in the set of data above, 32 has a frequency of 2.

Do-Anytime Activities

To work with your child on the concepts taught in this unit and in previous units, try these interesting and rewarding activities:

1. If you have a calculator at home, practice making (and breaking) tens.

   For example:

   Making tens: Enter 33. What needs to be done to display 50? $33 +$ _____ $= 50$

   Breaking tens: Enter 60. What needs to be done to display 52? $60 -$ _____ $= 52$

   Or, for more challenging practice, try the following:

   Enter 27. What needs to be done to display 40?

   Enter 90. What needs to be done to display 66?

   Try other similar numbers.

2. Make a game out of doubling, tripling, and quadrupling small numbers. For example, using the number 2, first double it. What number do you get? Continue the doubling process five more times. Start again with the number 2 and triple it; then quadruple it. Discuss the differences among the final numbers.

3. Collect a simple set of data from family and friends. For example, how high can different people's fingertips reach while the people are standing flat on the floor? Order the data to find the median.

Building Skills through Games

In Unit 7, your child will practice skills related to addition and subtraction as well as chance and probability by playing the following games:

Array Bingo

Players roll the dice and find an *Array Bingo* card with the same number of dots. Players then turn that card over. The first player to have a row, column, or diagonal of facedown cards calls out "Bingo!" and wins the game.

Soccer Spin

Players choose a spinner to use during the game. They choose a team to cheer for, Checks or Stripes. They then predict which team will win based on their spinner. They spin the spinner to check their prediction.

Basketball Addition

The basketball game is played by two teams, each consisting of 3–5 players. The number of points scored by each player in each half is determined by rolling a twenty-sided polyhedral die or 3 regular dice and using their sum. The team that scores the greater number of points wins the game.

Hit the Target

Players choose a 2-digit multiple of ten (10, 20, 30, and so on) as a target number. One player enters a starting number on a calculator and tries to change the starting number to the target number by adding a number to it on the calculator. Children practice finding differences between 2-digit numbers and higher multiples of tens.

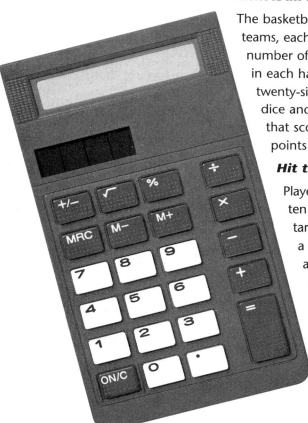

As You Help Your Child with Homework

As your child brings home assignments, you might want to go over the instructions together, clarifying them as necessary. The answers listed below will guide you through this unit's Home Links.

Home Link 7·1

1. 202, 204, 206, 208, 210, 212, 214, 216, 218

2. 500, 505, 510, 515, 520, 525, 530, 535, 540, 545

3. 550, 560, 570, 580, 590, 600, 610, 620, 630, 640

Home Link 7·2

1. 6; 7; 5; 9; 2   2. 6; 7; 5; 9; 8

3. 32 + 38; 65 + 5; 10 + 60; 43 + 27; 19 + 51; 51 + 19; 27 + 43

Home Link 7·3

1. Team A: 35; Team B: 25; A

2. Team A: 30; Team B: 35; B

3. Team A: 29; Team B: 40; B

4. Team A: 45; Team B: 59; B

Home Link 7·4

1.

in	out
12	6
50	25
40	20
30	15
16	8
18	9

Rule: Halve

2. 1, 2, 4, 8, 16, 32, 64

3. 3, 6, 12, 24, 48, 96, 192

4. 127 pennies, or $1.27

5. 9   6. 32   7. 38

Home Link 7·5

1. 8 pounds   2. 20 pounds   3. 5 pounds

4. 11,000 pounds   5. 199   6. 49

7. 107   8. 90

Home Link 7·6

5. 42   6. 103   7. 25   8. 29

Home Link 7·7

1. 3 points   7 points   9 points   (12 points)   15 points   20 points   21 points

2. 56 in.   66 in.   (68 in.)   70 in.   73 in.

3. 142 cm   168 cm   (173 cm)   178 cm   185 cm

Home Link 7·8

1. 2   2. 0   3. 46   4. 52

5. 9   6. 48   7. 49

HOME LINK 7·1 | Count by 2s, 5s, and 10s

Family Note In this lesson, your child has been counting by 2s, 5s, and 10s. After your child has completed these problems, help him or her look for patterns in the ones digits of the answers. In the example, the ones digits repeat: 0, 2, 4, 6, 8, 0, 2, 4, and so on. If your child is successful with these problems, ask him or her to count backward by 2s, 5s, or 10s. Start from a number that is a multiple of 10, such as 200.

Please return this Home Link to school tomorrow.

MRB 96

Example:

Count by 2s. Begin at 100. Write your first 10 counts below.

_100_, _102_, _104_, _106_, _108_, _110_, _112_, _114_, _116_, _118_

1. Count by 2s. Begin at 200. Write your first 10 counts below.

_200_, _____, _____, _____, _____, _____, _____, _____, _____, _____

2. Count by 5s. Begin at 500. Write your first 10 counts below.

_____, _____, _____, _____, _____, _____, _____, _____, _____, _____

3. Count by 10s. Begin at 550. Write your first 10 counts below.

_____, _____, _____, _____, _____, _____, _____, _____, _____, _____

Look at your counts. Write about any patterns you find in the counts.

_____

_____

_____

Missing Addends

Family Note In this lesson, your child found the difference between a number and a multiple of 10. In Problems 1 and 2, your child will find the difference between a number and the next-higher multiple of 10. For example, your child will determine which number added to 62 equals 70 (8). In Problem 3, your child will find different combinations of numbers that add to 70. If your child has difficulty with this problem, suggest changing the first number in each combination to the next-higher multiple of 10. For example, add 2 to 48 to make 50 and then add 20 to 50 to make 70. 2 + 20 = 22, so 48 + 22 = 70.

Please return this Home Link to school tomorrow.

1. 4 + _____ = 10

   10 = 3 + _____

   _____ + 5 = 10

   10 = _____ + 1

   8 + _____ = 10

2. 54 + _____ = 60

   90 = 83 + _____

   75 + _____ = 80

   40 = 31 + _____

   _____ + 62 = 70

Unit

3. Make 70s. Show someone at home how you did it.

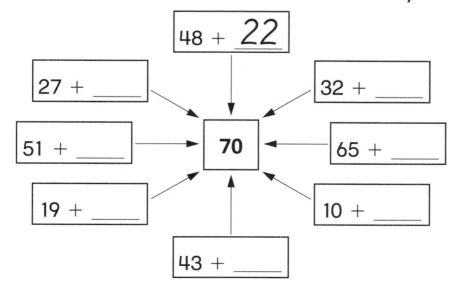

48 + __22__

27 + _____

32 + _____

51 + _____   **70**   65 + _____

19 + _____

10 + _____

43 + _____

135

HOME LINK 7·3 Who Scored More Points?

> **Family Note** In this lesson, your child added three or more 1-digit and 2-digit numbers. As your child completes the problems below, encourage him or her to share the different ways in which the points can be added. Your child might add all the tens first and then add all the ones. For example, $20 + 5 + 4 + 6 = 20 + 15 = 35$. Your child may also look for combinations of numbers that are easier to add. In Game 1, for example, first add 14 and 6 to get 20 and then add 15 to get 35.
>
> *Please return this Home Link to school tomorrow.*

Do the following for each problem:

Unit
points

➜ Add the points for each team.

➜ Decide which team scored more points. The team with the greater number of points wins the game.

➜ Circle your answer.

1. Game 1

Team A:
$15 + 14 + 6 =$ _____

Team B:
$5 + 13 + 7 =$ _____

Who won? A or B

2. Game 2

Team A:
$12 + 6 + 4 + 8 =$ _____

Team B:
$5 + 10 + 19 + 1 =$ _____

Who won? A or B

3. Game 3

Team A:
$17 + 4 + 5 + 3 =$ _____

Team B:
$2 + 11 + 9 + 18 =$ _____

Who won? A or B

4. Game 4

Team A:
$7 + 4 + 16 + 13 + 5 =$ _____

Team B:
$22 + 9 + 8 + 3 + 17 =$ _____

Who won? A or B

137

HOME LINK 7·4 Doubles and Halves

Family Note In today's lesson, your child heard a story and used a calculator to double numbers and find halves of numbers repeatedly. Help your child solve the doubling and halving problems below. When appropriate, have your child use money or counters to help solve the problems. In Problem 1, for example, your child might display 40 counters, divide them into two equal groups, and then count to find that half of 40 is 20.

Please return this Home Link to school tomorrow.

1. Write a rule in the rule box. Then complete the table.

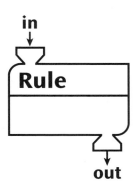

in	out
12	6
50	25
40	
30	
	8
	9

2. Fill in the frames using the rule in the rule box.

Rule
Double

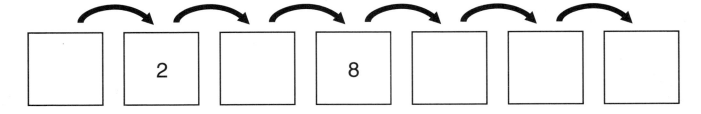

| | 2 | | 8 | | | |

HOME LINK 7·4 | **Doubles and Halves** *continued*

3. Fill in the frames using the rule in the rule box.

| **Rule** |
| Double |

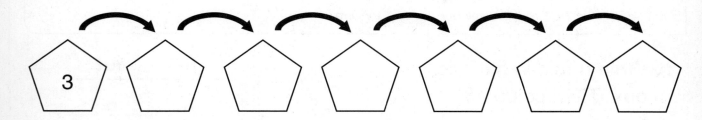

| 3 |

Try This

4. Maria finds 1 penny under her pillow when she wakes up on Monday morning. On Tuesday, she finds 2 pennies. On Wednesday, she finds 4 pennies. So, on Wednesday, she has a total of 7 cents.

On Thursday, Friday, Saturday, and Sunday, Maria finds double the amount of money she found under her pillow the day before. How much money does Maria have on Sunday?

_____

Practice

5. $28 - 19 = $ _____

6. $74 - 42 = $ _____

7. $67 - 29 = $ _____

| **Unit** |
| |

HOME LINK 7·5 Estimating Weights

Family Note In today's lesson, your child practiced reading weights, in pounds, on a bath scale. One purpose of this activity is to improve your child's perception of weight so he or she can make more realistic estimates of weights. To help develop your child's ability to read a bath scale, take every opportunity at home to use your bath scale to determine the weights of objects.

Please return this Home Link to school tomorrow.

Circle the best estimate for the weight of each object.

1. newborn baby

8 pounds

20 pounds

70 pounds

2. Thanksgiving turkey

$\frac{1}{2}$ pound

20 pounds

70 pounds

3. bag of apples

5 pounds

35 pounds

65 pounds

4. An adult bull African elephant (the largest animal on land)

100 pounds

500 pounds

11,000 pounds

Practice

5. 236
      − 37

6. 199
      − 150

7. 78
      + 29

8. 45
      + 45

HOME LINK 7·6

Comparing Arm Spans

Family Note In today's lesson, your child measured his or her standing long jump in centimeters and his or her arm span in inches. Help your child compare his or her arm span to someone else's arm span at home. Also, help your child find objects in the house that are about the same length as his or her arm span.

Please return this Home Link to school tomorrow.

MRB
62

My arm span is about _____ inches long.

1. Tell someone at home about how long your arm span is in inches.

2. Compare your arm span to someone at home. Can you find someone who has a longer arm span than you do? Is there someone at home who has a shorter arm span?

   _____ has a longer arm span than I do.

   _____ has a shorter arm span than I do.

3. List some objects below that are about the same length as your arm span.

   _____   _____

   _____   _____

4. Explain how you know the objects you listed in Problem 3 are about the same length as your arm span.

   _____

   _____

   _____

Practice

5.	23	6.	45	7.	64	8.	86
	+ 19		+ 58		− 39		− 57

Find the Middle Value

> **Family Note**
>
> In this lesson, your child sorted data to find the median. *Median* is a term used for the middle value. To find the median of a set of data, arrange the data in order from smallest to largest. Count from either end to the number in the middle. The middle value is the median. As your child finds the median in Problems 2 and 3, remind him or her that "in." is the abbreviation for inches and "cm" is the abbreviation for centimeters.
>
> *Please return this Home Link to school tomorrow.*

MRB 46

List the data in order from smallest to largest.

Draw a circle around the median in your list.

1.

12 points	3 points	21 points	15 points	20 points	7 points	9 points

_____   _____   _____   _____   _____   _____   _____
points   points   points   points   points   points   points
smallest   **largest**

HOME LINK 7·7

Find the Middle Value *continued*

2.

Jarel: 66 in. tall	Suki: 70 in. tall	Peter: 56 in. tall	Keisha: 73 in. tall	Cesar: 68 in. tall

_____ in.   _____ in.   _____ in.   _____ in.   _____ in.

smallest   **largest**

3.

Jarel: 168 cm tall	Suki: 178 cm tall	Peter: 142 cm tall	Keisha: 185 cm tall	Cesar: 173 cm tall

_____ cm   _____ cm   _____ cm   _____ cm   _____ cm

smallest   **largest**

HOME LINK 7·8 | Interpreting Data

Ms. Ortiz is a basketball coach. She measured the height of each player. Then she made the data table shown below.

1. How many players are

50 inches tall? _____ players

2. How many players are

47 inches tall? _____ players

3. The shortest player is

_____ inches tall.

4. The tallest player is

_____ inches tall.

5. How many players did Ms. Ortiz

measure? _____ players

6. Which height occurs most often? _____ inches

7. Find the middle (median) height. _____ inches

Players' Heights	
Height (inches)	Number of Players
46	1
47	0
48	3
49	1
50	2
51	1
52	1

HOME LINK 7·9   # Unit 8: Family Letter

Fractions

In Unit 8, children will review and extend concepts of fractions. Specifically, they will recognize fractions as names for parts of a whole, or ONE.

Children will see that, as with whole numbers, many different fractions can name the same quantity. For example, $\frac{2}{4}$ and $\frac{6}{12}$ are names for $\frac{1}{2}$.

Children will also explore relationships among fractions as they work with pattern-block shapes and Fraction Cards that show shaded regions.

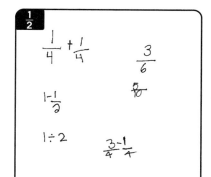

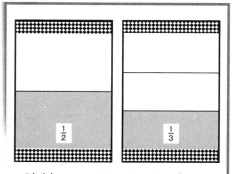

Children use Fraction Cards to compare fractions by looking at the shaded areas.

Please keep this Family Letter for reference as your child works through Unit 8.

Vocabulary

Important terms in Unit 8:

fraction A number that names equal parts of a whole, or ONE.

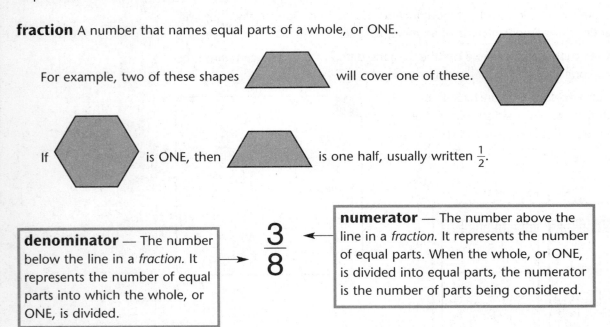

For example, two of these shapes will cover one of these.

If is ONE, then is one half, usually written $\frac{1}{2}$.

denominator — The number below the line in a *fraction*. It represents the number of equal parts into which the whole, or ONE, is divided.

$$\frac{3}{8}$$

numerator — The number above the line in a *fraction*. It represents the number of equal parts. When the whole, or ONE, is divided into equal parts, the numerator is the number of parts being considered.

It is not necessary for children to use the words numerator and denominator now. They will learn them over time with repeated exposure. Do, however, use these words, as well as the informal "number on the top" and "number on the bottom," when you discuss fractions with your child.

equivalent fractions *Fractions* with different denominators that name the same number. For example, $\frac{1}{2}$ and $\frac{2}{4}$ are equivalent fractions.

Do-Anytime Activities

To work with your child on the concepts taught in this unit and in previous units, try these interesting and rewarding activities:

1. Review fraction notation. For example, ask: "In a fraction, what does the number on the bottom (the denominator) tell you?" "What does the number on the top (the numerator) tell you?"

2. Draw a picture of a rectangular cake, a circular pizza, or a similar food (better yet, have the real thing). Discuss ways to cut the food to feed various numbers of people so each person gets an equal portion.

3. Read a recipe and discuss the fractions in it. For example, ask: "How many $\frac{1}{4}$ cups of sugar would we need to get 1 cup of sugar?"

4. Compare two fractions and tell which is larger. For example, ask: "Which would give you more of a pizza: $\frac{1}{8}$ of it, or $\frac{1}{4}$?"

As You Help Your Child with Homework

As your child brings home assignments, you might want to go over the instructions together, clarifying them as necessary. The answers listed below will guide you through this unit's Home Links.

Home Link 8·1

1. $\frac{1}{2}$; $\frac{1}{2}$ 2. $\frac{3}{4}$; $\frac{1}{4}$

Home Link 8·2

1. $\frac{1}{2}$ 2. $\frac{1}{6}$ 3. $\frac{2}{3}$ 4. 101

5. 101   6. 132   7. 158

Home Link 8·3

1. 4; 4; 8   2. 44   3. 98

4. 38   5. 90

Home Link 8·4

1. $\frac{1}{2} = \frac{2}{4}$ 2. $\frac{1}{2} = \frac{4}{8}$ 3. $\frac{1}{4} = \frac{4}{16}$

4. $\frac{1}{4} = \frac{2}{8}$ 5. $\frac{1}{5} = \frac{4}{20}$ 6. 100

7. 82

Home Link 8·5

1.   2.

3.   4. 84   5. 133

Home Link 8·6

1. Answers vary.   2. Answers vary.

3. 77   4. 37   5. 94   6. 15

Home Link 8·7

1. $\frac{4}{7}$ 2. $\frac{2}{12}$, or $\frac{1}{6}$ 3. $\frac{1}{3}$ 4. 4 tulips

5. 104   6. 53   7. 21   8. 39

Building Skills through Games

In Unit 8, your child will practice multiplication and fraction skills by playing the following games:

Array Bingo

Players roll the dice and find an *Array Bingo* card with the same number of dots. Players then turn that card over. The first player to have a row, column, or diagonal of facedown cards calls "Bingo!" and wins the game.

Equivalent Fractions Game

Players take turns turning over Fraction Cards and finding matching cards that show equivalent fractions.

Fraction Top-It

Players turn over two Fraction Cards and compare the shaded parts of the cards. The player with the larger fraction keeps both of the cards. The player with more cards at the end wins.

Name That Number

Each player turns over a card to find a number that must be renamed using any combination of five faceup cards.

HOME LINK 8·1 | **Equal Parts**

Family Note Help your child collect things that can be easily folded into equal parts. As your child works with fractions, remind him or her that the number under the fraction bar, the *denominator*, gives the total number of equal parts into which the whole is divided. The number over the fraction bar, the *numerator*, tells the number of equal parts that are being considered. Don't expect your child to use these words. They will be learned over time with repeated exposure.

Please return this Home Link to school tomorrow.

MRB 12

Use a straightedge.

1. Divide the shape into 2 equal parts. Color 1 part.

   Part colored = ⬚/⬚ Part not colored = ⬚/⬚

   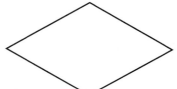

2. Divide the shape into 4 equal parts. Color 3 parts.

   Part colored = ⬚/⬚ Part not colored = ⬚/⬚

   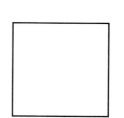

3. Fold some things into equal parts.

   Examples: paper napkin, paper plate, magazine picture

   Label each part with a fraction. Show your folded things to someone at home. Talk about what the fractions mean.

   Bring the things you folded to school for the Fractions Museum.

   I folded a _____ into _____ equal parts.

   Each part shows _____.

Fractions of Shapes

> **Family Note** In today's lesson, your child compared pattern blocks, one of which represents ONE whole. As you work on this activity with your child, keep in mind that the shape below the ONE is a fractional part of the whole shape. Remind your child that the size of a fractional part of a whole depends on the size of the whole.
>
> *Please return this Home Link to school tomorrow.*

1. If this shape is ONE,

then ⟋ is what fraction of the shape? _____

2. If this shape is ONE, ⬡

then ▽ is what fraction of the shape? _____

Try This

3. If this shape is ONE, ⬭

then ▱ is what fraction of the whole shape? _____

Practice

Solve.

	Unit

4.   75
     + 26

5.   56
     + 45

6.   84
     + 48

7.   91
     + 67

HOME LINK 3·3 | # Fractions of Collections

Family Note In this lesson, your child learned to use fractions to name part of a collection of objects. For example, your child could identify 2 out of 4 objects as $\frac{2}{4}$ or $\frac{1}{2}$. Show your child how to use pennies to act out Problem 1. Help your child collect household items that can be separated into fractional parts—or any other items that have fractions written on them. Encourage your child to bring these items to school for the class's Fractions Museum.

Please return this Home Link to school tomorrow.

1.

Three people share 12 pennies. Circle each person's share.

How many pennies does each person get? _____ pennies

$\frac{1}{3}$ of 12 pennies = _____ pennies.

$\frac{2}{3}$ of 12 pennies = _____ pennies.

Practice

Solve.

		Unit
		Cars

2. 68 − 24 = _____   **4.**   65   **5.**   64

                                  − 27   + 26

3. 53 + 45 = _____   ‾‾‾‾   ‾‾‾‾

Ask someone at home to help you find more things to bring to school for the Fractions Museum.

HOME LINK **B·4** | # Shading Fractional Parts

Family Note In this lesson, your child learned that a fractional part of a whole can be named in many different ways with *equivalent* fractions. For example, $\frac{2}{4}$, $\frac{4}{8}$, and $\frac{3}{6}$ are names for $\frac{1}{2}$, while $\frac{2}{8}$ and $\frac{4}{16}$ are names for $\frac{1}{4}$. Help your child shade each of the shapes below to show the appropriate fraction. Make sure your child understands that the fractions are equivalent because they name the same part of the shape.

Please return this Home Link to school tomorrow.

1. Shade $\frac{1}{2}$ of the rectangle.

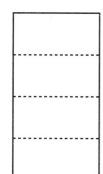 $\frac{1}{2} = \frac{\Box}{4}$

2. Shade $\frac{1}{2}$ of the rectangle.

$\frac{1}{2} = \frac{\Box}{8}$

3. Shade $\frac{1}{4}$ of the square.

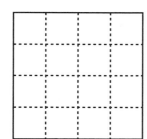 $\frac{1}{4} = \frac{\Box}{16}$

4. Shade $\frac{1}{4}$ of the square.

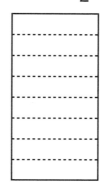 $\frac{1}{4} = \frac{\Box}{8}$

Try This

5. Shade $\frac{1}{5}$ of the rectangle.

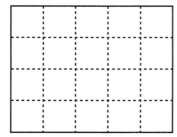 $\frac{1}{5} = \frac{\Box}{20}$

Practice

Solve.

6. $130 - 30 =$ _____

7. $37 + 45 =$ _____

159

HOME LINK
8·5

Fractions of Regions

> **Family Note** In today's lesson, your child played a game in which he or she matched pictures of equivalent fractions. Stress the idea to your child that equivalent fractions show different ways to name a fractional part of a whole.
>
> *Please return this Home Link to school tomorrow.*

1. Circle the pictures that show $\frac{1}{2}$ of the rectangle shaded.

 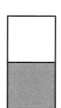  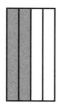

2. Circle the pictures that show $\frac{3}{4}$ of the rectangle shaded.

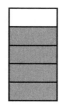  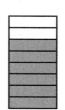 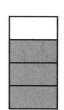

3. Circle the pictures that show $\frac{2}{3}$ of the rectangle shaded.

 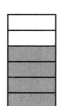 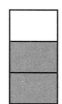 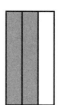

Practice

Add.

Unit

4. $\begin{array}{r} 36 \\ + 48 \\ \hline \end{array}$

5. $\begin{array}{r} 76 \\ + 57 \\ \hline \end{array}$

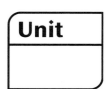
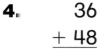

Name _____ Date _____ Time _____

More or Less Than $\frac{1}{2}$?

> **Family Note** The class has begun to compare fractions by identifying fractions that are less than, more than, or equivalent to $\frac{1}{2}$. One way that children can compare fractions is if the numerator is less than half of the denominator, then the fraction is less than $\frac{1}{2}$; if the numerator is more than half of the denominator, then the fraction is greater than $\frac{1}{2}$; if the numerator is exactly half of the denominator, then the fraction is the same as, or equivalent to, $\frac{1}{2}$. To do the activity below, ask your child to draw a fraction bar on a sheet of paper. Your child might want to use the bar to create fractions by positioning the number tiles above and below the bar.
>
> *Please return this Home Link to school tomorrow.*

Cut out the number tiles.

1. Use the tiles to make fractions that are less than $\frac{1}{2}$.

Make as many fractions as you can. **Example:** $\frac{1}{3}$
Record the fractions you make.

_____

2. Use the tiles to make fractions that are more than $\frac{1}{2}$.

Make as many fractions as you can. **Example:** $\frac{2}{3}$
Record the fractions you make.

_____

_____

| 0 |
| 1 |
| 2 |
| 3 |
| 4 |
| 5 |
| 6 |
| 7 |
| 8 |
| 9 |

Practice

Solve.

Unit

3. $23 + 54 =$ _____

4. $73 - 36 =$ _____

5. $\begin{array}{r} 56 \\ + 38 \\ \hline \end{array}$

6. $\begin{array}{r} 42 \\ - 27 \\ \hline \end{array}$

HOME LINK B·7 | **Fractions**

| **Family Note** | In this lesson, your child has been completing number stories about fractions. Encourage your child to draw pictures or use small objects, such as pennies, to help him or her complete fraction number stories. |

Please return this Home Link to school tomorrow.

1. 7 children are waiting for the school bus.

   4 of them are girls.

   What fraction of the children are girls? _____

2. 12 dogs were in the park.

   2 of them were dalmatians.

   What fraction of the dogs were dalmatians? _____

3. There are 15 cupcakes.

   5 of the cupcakes are chocolate.

   What fraction of the cupcakes are chocolate? _____

4. There are 16 tulips in the garden.

   $\frac{1}{4}$ of the tulips are red.

   How many tulips are red? _____ tulips

Practice

Unit [ ]

Solve.

5.	23	6.	17	7.	42	8.	78
	+ 81		+ 36		− 21		− 39

HOME LINK 8·8

Unit 9: Family Letter

Measurement

In Unit 9, children will explore measurements of various types. Your child will be asked to look for examples of measurements and measuring tools to bring to school for the Measures All Around Museum. The examples will help children appreciate the important role that measurement plays in everyday life.

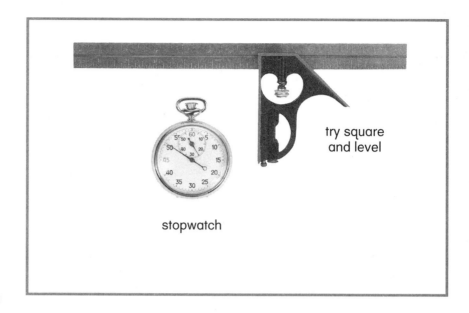

stopwatch

try square and level

Children will estimate and measure distances by inch, foot, and yard, as well as centimeter, decimeter, and meter. Children will learn that measurements are not always exact; they will use terms such as *close to, between,* and *about* when describing measurements. For closer or more exact measurements, children will measure to the nearest half-inch and half-centimeter.

In addition to measures of length, children will explore the areas of shapes using square inches and square centimeters. Children will also begin to develop a sense of the size of units of capacity and weight, such as cups and liters and pounds and kilograms.

Everyday Mathematics uses U.S. customary and metric units of measure. Although children make conversions within each system (length, capacity, or weight), they will not make conversions from one system to the other at this time.

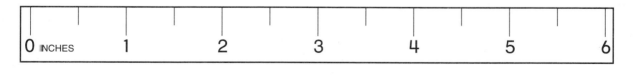

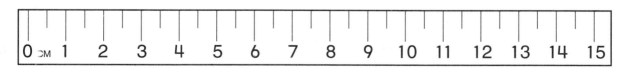

Please keep this Family Letter for reference as your child works through Unit 9.

Vocabulary

Important terms in Unit 9:

capacity The amount a container can hold. The volume of a container. Capacity is usually measured in units such as gallons, pints, cups, fluid ounces, liters, and milliliters.

perimeter The distance around a 2-dimensional shape, along the boundary of the shape. (The perimeter measures the length of a shape's "rim.")

area The amount of surface inside a 2-dimensional figure. Area is measured in square units, such as square inches or square centimeters.

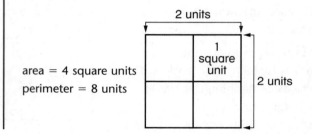

area = 4 square units
perimeter = 8 units

Metric System

Units of Length

1 meter (m)	= 10 decimeters (dm)
	= 100 centimeters (cm)
1 decimeter	= 10 centimeters
1 kilometer (km)	= 1,000 meters

Units of Weight

1 kilogram (kg)	= 1,000 grams (g)

Units of Capacity

1 liter (L)	= 1,000 milliliters (mL)
$\frac{1}{2}$ liter	= 500 milliliters

U.S. Customary System

Units of Length

1 yard (yd)	= 3 feet (ft)
	= 36 inches (in.)
1 foot	= 12 inches
1 mile (mi)	= 1,760 yards
	= 5,280 feet

Units of Weight

1 pound (lb)	= 16 ounces (oz)
2,000 pounds	= 1 ton (T)

Units of Capacity

1 cup (c)	= $\frac{1}{2}$ pint (pt)
1 pint	= 2 cups
1 quart (qt)	= 2 pints
1 half-gallon $\left(\frac{1}{2}\ \text{gal}\right)$	= 2 quarts
1 gallon (gal)	= 4 quarts

Do-Anytime Activities

To work with your child on the concepts taught in this unit and in previous units, try these interesting and rewarding activities:

1. Gather a tape measure, a yardstick, a ruler, a cup, a gallon container, and a scale. Discuss the various things you and your child can measure— for example, the length of a room, how many cups are needed to fill a gallon container, and your child's weight alone and when he or she is holding objects such as books. Record the data and continue to measure and weigh different items periodically.

2. Mark certain routes on a road map and together figure the distance between two points in miles and kilometers.

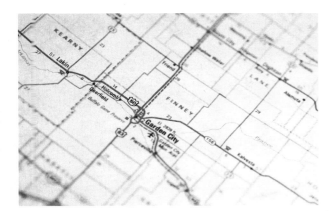

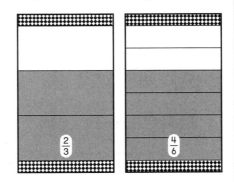

Building Skills through Games

In Unit 9, your child will practice mathematical skills by playing the following games:

Equivalent Fractions Game

Players take turns turning over Fraction Cards and try to find matching cards that show equivalent fractions.

Fraction Top-It

Players turn over two Fraction Cards and compare the shaded parts of the cards. The player with the larger fraction keeps both cards. The player with more cards at the end wins!

$\frac{2}{3}$ $\frac{4}{6}$

Name That Number

Each player turns over a card to find a number that must be renamed using any combination of five faceup cards.

Number-Grid Difference Game

Players subtract 2-digit numbers using the number grid.

As You Help Your Child with Homework

As your child brings home assignments, you may want to go over the instructions together, clarifying them as necessary. The answers listed below will guide you through this unit's Home Links.

Home Link 9·1

5. 115   **6.** 791

7. 46   **8.** 325

Home Link 9·2

3. 12 inches   **4.** 3 feet

5. 10 centimeters   **6.** 100 centimeters

7. 24 inches   **8.** 9 feet

9. 40 centimeters   **10.** 700 centimeters

11. 69   **12.** 85

13. 48   **14.** 37

Home Link 9·3

1. $2\frac{1}{2}$ inches   **2.** 4 inches

3. 3 centimeters   **4.** 9 centimeters

10. 290   **11.** 397

Home Link 9·4

1. Perimeter: 6 or 7 inches

2. Perimeter: $4\frac{1}{2}$ or 5 inches

3. Answer: 47 feet. Sample number models:

$14 + 14 + 9\frac{1}{2} + 9\frac{1}{2} = 47$ or

$2 \times 14 + 2 \times 9\frac{1}{2} = 47$

Home Link 9·5

1. 214   **2.** 113

Home Link 9·6

2.

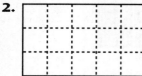
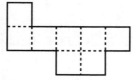

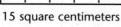 15 square centimeters   8 square centimeters

3. 359   **4.** 794

5. 400   **6.** 401

Home Link 9·7

1. 9 sq cm   **2.** 11 sq cm

3. 10 sq cm   **4.** I: 20 cm

                      U: 24 cm

                      J: 22 cm

5. 95   **6.** 92

7. 162   **8.** 103

Home Link 9·8

Rule
1 gal = 4 qt

gal	qt
2	8
4	16
6	24
10	40

Answers vary.

1. 83   **2.** 34

Home Link 9·9

1. 159   **2.** 177

170

Using Measurement

> **Family Note** In class today, your child measured distances by using a yardstick. Talk with your child about measurements that you use at your job, around the house, in sports, or in other activities. If you don't have measuring tools to show your child, you might find pictures of measuring tools in a catalog, magazine, or book. Discuss with your child how these tools are used.
>
> *Please return this Home Link to school tomorrow.*
>
> MRB 64–66

1. Talk with people at home about how they use measurements at home, at their jobs, or in other activities.

2. Ask people at home to show you the tools they use for measuring. Write the names of some of these tools. Be ready to talk about your list in class.

_____   _____

_____   _____

_____   _____

3. Look for measures in pictures in newspapers or magazines. For example, an ad might name the height of a bookcase or tell how much a container holds. Ask an adult if you may bring the pictures to school for our Measures All Around Museum. Circle the measures.

4. Bring one or two small boxes shaped like rectangular prisms to school. The boxes should be small enough to fit on a sheet of paper. **You will need these for Lesson 9-4.**

Practice

5. 86
  + 29

6. 770
  + 21

7. 60
  − 14

8. 350
  − 25

HOME LINK 9·2

Linear Measurements

Family Note

Today your child reviewed how to use a ruler to measure objects and distances in inches and feet and in centimeters and decimeters. Your child's class also began making a Table of Equivalent Measures for the U.S. customary and metric systems. Ask your child to show you how to measure some of the objects or distances that he or she selects to complete the tables below.

Please return this Home Link to school tomorrow.

MRB
64–67

1. Cut out the 6-inch ruler on the next page. Measure two objects or distances. Measure to the nearest foot. Then measure again to the nearest inch. Some things you might measure are the width of the refrigerator door, the length of the bathtub, or the height of a light switch from the floor.

Object *or* Distance	Nearest Foot	Nearest Inch
	about _____ ft	about _____ in.
	about _____ ft	about _____ in.

2. Cut out the 10-centimeter ruler on the next page. Measure the same objects or distances. Measure to the nearest decimeter. Then measure again to the nearest centimeter.

Object *or* Distance	Nearest Decimeter	Nearest Centimeter
	about _____ dm	about _____ cm
	about _____ dm	about _____ cm

HOME LINK 9·2 | **Linear Measurements** *continued*

Complete each sentence.

3. One foot is equal to _____ inches.

4. One yard is equal to _____ feet.

5. One decimeter is equal to _____ centimeters.

6. One meter is equal to _____ centimeters.

7. Two feet are equal to _____ inches.

8. Three yards are equal to _____ feet.

9. Four decimeters are equal to _____ centimeters.

10. Seven meters are equal to _____ centimeters.

___

Practice

11. $23 + 46 =$ _____

12. $38 + 47 =$ _____

13. $84 - 36 =$ _____

14. $76 - 39 =$ _____

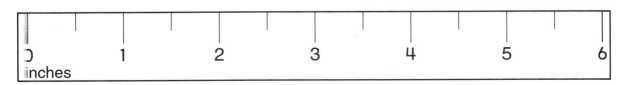

Measuring Lengths

Family Note Today your child measured life-size pictures of objects to the nearest half-inch and half-centimeter. Take turns with your child measuring objects to the nearest half-inch and half-centimeter. Check to see if your measurements are the same.

Please return this Home Link to school tomorrow.

MRB
64–66

t out the 6-inch ruler on the next page. Measure each line segment the nearest half-inch. Write the measurement in the blank to the ht o⁻ each segment.

. _____ _____ inches

. _____ _____ inches

t out the 15-centimeter ruler on the next page. Measure each line gment to the nearest half-centimeter. Write the measurement in the nk to the right of each segment.

. _____ _____ centimeters

. _____ _____ centimeters

asu⁻e some objects in your home to the nearest half-inch or f-centimeter. List the objects and their measurements below.

Object	Measurement
_____	_____
_____	_____
_____	_____
_____	_____

HOME LINK 9·3 | **Measuring Lengths** *continued*

9. Draw pictures of two things you measured. Mark the parts you measured. Record the measurements under the pictures.

Practice

10. 231
　　+ 59

11. 452
　　− 55

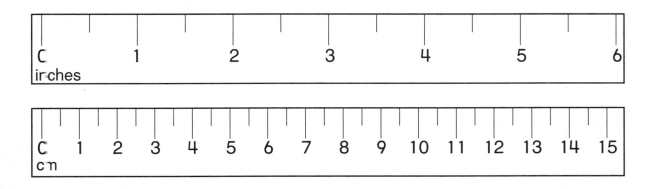

HOME LINK 9·4 Perimeter

Family Note Today your child found the perimeter of different shapes and the distance around his or her thumb, wrist, neck, and ankle. Perimeter is the measure around something. Finding perimeters also gives your child practice in measuring to the nearest inch and centimeter.

Please return this Home Link to school tomorrow.

Cut out the 6-inch ruler on the next page. Measure the side of each figure to the nearest inch. Write the length next to each side. Then find the perimeter.

1.

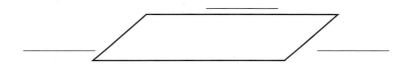

Perimeter: _____ inches

2

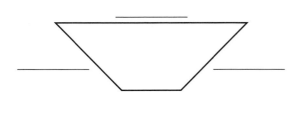

Perimeter: _____ inches

181

ME LINK
9·4

Perimeter *continued*

lve the number story. Write a number model.

. Mr. Lopez is putting a fence around his vegetable garden.
The garden is shaped liked a rectangle. The longer sides
are 14 feet long, and the shorter sides are $9\frac{1}{2}$ feet long.
How much fencing should Mr. Lopez buy?

Answer: _____ feet

Number model: _____

. Draw a quadrangle below. Measure the sides to the
nearest $\frac{1}{2}$-inch. Write the length next to each side.
Find the perimeter.

The perimeter of my quadrangle is _____ inches.

0		1		2		3		4		5		6
inches												

HOME LINK 9·5 | Travel Interview

Family Note Our class is studying measurement of longer distances. If the traveler your child talks to had experiences with the metric system in another country, have your child include this information to share with the class.

Please return this Home Link to school tomorrow.

Ask someone at home to tell you about the longest trip he or she ever took. Write about the trip. Here are some questions you might want to ask that person:

When did you take the trip?

Where did you go?

What interesting or unusual things did you see or do?

How did you travel? By car? By plane? By train?

How long did the trip take?

How far did you travel?

_____

_____

_____

_____

_____

_____

_____

_____

Practice

1. 136 + 78 = _____

2. 172 − 59 = _____

HOME LINK 9·6 **Capacity and Area**

> **Family Note** Today your child explored the ideas of *capacity* and *area*. Before your child is exposed to formal work with these measures (such as equivalent units of capacity or formulas for finding area), it is important that he or she have an informal understanding of these measures.
>
> In Problem 1, help your child see that although the glasses may have different dimensions, they can still hold about the same amount of water. In Problem 2, the number of squares that your child counts is the area in square centimeters.
>
> *Please return this Home Link to school tomorrow.*

1. Find two different glasses at home that you think hold about the same amount of water. Test your prediction by pouring water from one glass into the other. Do they hold about the same amount of water? Does one glass hold more than the other? Explain to someone at home how you know.

2. Count squares to find the area of each figure.

_____ square centimeters   _____ square centimeters

Practice

3. $459 - 100 =$ _____

4. $594 + 200 =$ _____

5. $\begin{array}{r} 350 \\ +\ 50 \\ \hline \end{array}$

6. $\begin{array}{r} 460 \\ -\ 59 \\ \hline \end{array}$

HOME LINK 9·7 Area and Perimeter

Family Note Today children discussed the concept of finding the area of a surface. Area is measured by finding the number of square units needed to cover the surface inside a shape. Make sure your child understands that, when he or she is finding the perimeter of the letters in Problem 4, he or she is finding the distance around the outside of the letters.

Please return this Home Link to school tomorrow.

MRB
69

Find the area of each letter.

1.

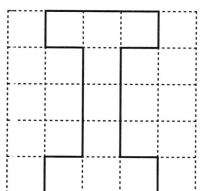

2.

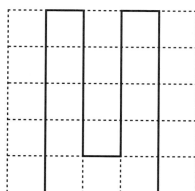

3.

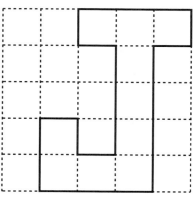

Area = _____ sq cm   Area = _____ sq cm   Area = _____ sq cm

4. What is the perimeter of each letter?

I: _____ cm   U: _____ cm   J: _____ cm

Practice

5. 67 + 28 = _____

6. 154 − 62 = _____

7.   88
      + 74
   _____

8.   126
       − 23
    _____

HOME LINK 9·8 Capacity

> **Family Note**
> Today children discussed units of capacity. Capacity is a measure of the amount of space something occupies or contains. Your child recorded equivalent U.S. customary units of capacity (cup, pint, quart, half-gallon, gallon) and equivalent metric units of capacity (milliliter, liter). Please help your child pick out a recipe and identify the units of capacity in the list of ingredients.
>
> *Please return this Home Link to school tomorrow.*

Ask someone at home to help you find a recipe that uses units of capacity. Copy those ingredients and the amounts that are used in the recipe. Bring your list to school.

Example: $\frac{3}{4}$ *cup of milk*

_____

_____

_____

"What's My Rule?"

Rule
1 gal = 4 qt

gal	qt
2	
	16
6	
10	

Practice

1.   27
      + 56

2.   92
      − 58

HOME LINK 9·9 · Weight

> **Family Note** Today children discussed U.S. customary units of weight (pounds, ounces) and metric units of weight (grams, kilograms). Your child weighed different objects using a variety of scales. Help your child weigh items using scales in your home or find items with weights written on them.
>
> *Please return this Home Link to school tomorrow.*

Find out what kinds of scales you have at home—for example, a bath scale, a letter scale, or a package scale. Weigh a variety of things on the scales, such as a person, a letter, or a book. Record your results below.

If you don't have any scales, look for cans and packages of food with weights written on them. Record those weights below. Remember that ounces (oz) measure weight and that fluid ounces (fl oz) measure capacity.

Object	Weight (include unit)
_____	_____
_____	_____
_____	_____
_____	_____
_____	_____
_____	_____

Practice

1. 86 + 73 = _____   **2.** 132 + 45 = _____

Unit 10: Family Letter

Decimals and Place Value

In this unit, children will review money concepts, such as names of coins and bills, money exchanges, and equivalent amounts. They will pretend to pay for items and to make change.

The unit also focuses on extending work with fractions and money by using decimal notation. Children will use calculators for money problems and estimation.

Later in this unit, children will work with place-value notation for 5-digit numbers. Here, as previously, the focus remains on strategies that help children automatically think of any digit in a numeral in terms of its value as determined by its place. For example, children will learn that in a number like 7,843, the 8 stands for 800, not 8, and the 4 for 40, not 4.

50¢

50 cents

$\frac{1}{2}$ of a dollar

$0.50

fifty cents

Ⓓ Ⓓ Ⓓ Ⓓ Ⓓ

Please keep this Family Letter for reference as your child works through Unit 10.

Vocabulary

Important terms in Unit 10:

decimal point A mark used to separate the ones and tenths places in decimals. A decimal point separates dollars from cents in money notation. The mark is a dot in the U.S. customary system and a comma in Europe and some other countries.

flat In *Everyday Mathematics*, the base-10 block consisting of one hundred 1-centimeter cubes.

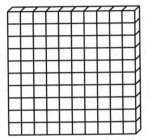

long In *Everyday Mathematics*, the base-10 block consisting of ten 1-centimeter cubes.

cube In *Everyday Mathematics*, the smaller cube of the base-10 blocks, measuring 1 centimeter on each edge.

place value A system that gives a digit a value according to its position in a number. In our standard *base-10* (decimal) system for writing numbers, each place has a value 10 times that of the place to its right and one-tenth the value of the place to its left. The chart below illustrates the place value of each digit in 7,843.

thousands	,	hundreds	tens	ones
7	,	8	4	3

Building Skills through Games

In Unit 10, your child will build his or her understanding of fractions and money by playing the following games:

Fraction Top-It

Players turn over two fraction cards and compare the shaded parts of the cards. The player with the larger fraction keeps both cards. The player with more cards wins.

Money Exchange Game

Players roll a die and put that number of $1 bills on their Place-Value Mats. Whenever possible, they exchange ten $1 bills for one $10 bill. The first player to make an exchange for one $100 bill wins.

Pick-a-Coin

Players create coin collections based on rolls of a die. Players try to get the largest possible values for their collections.

Spinning for Money

Players "spin the wheel" to find out which coins they will take from the bank. The first player to exchange his or her coins for a dollar wins.

Equivalent Fractions Game

Players take turns turning over Fraction Cards and try to find matching cards that show equivalent fractions.

Do-Anytime Activities

To work with your child on the concepts taught in this unit and in previous units, try these interesting and rewarding activities:

1. Collect a variety of coins and help your child count them. Discuss what other coin combinations would equal the same amount. For example, each group of coins shown on this page equals $1.00.

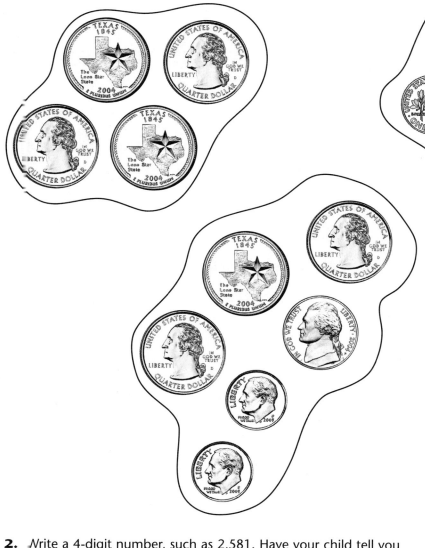

2. Write a 4-digit number, such as 2,581. Have your child tell you the place value of each digit. Rearrange the digits several times, pointing out the change in place value for each of the new number's digits. In 2,581, the 2 stands for 2,000; the 5, 500; the 8, 80; and the 1, 1.

3. Ask your child to add up grocery receipts by using a calculator.

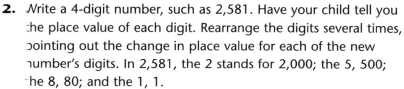

As You Help Your Child with Homework

As your child brings home assignments, you may want to go over the instructions together, clarifying them as necessary. The answers listed below will guide you through this unit's Home Links.

Home Link 10·1

1. 10 pennies = 10¢, or $0.10
10 nickels = 50¢, or $0.50
10 dimes = $1.00
10 quarters = $2.50
10 half-dollars = $5.00
Total = $9.10

Home Link 10·2

1. $3.57   **2.** $3.55   **3.** $0.52   **4.** $0.08

5. Sample answers: $1 $1 Q Q D P P P P or
$1 Q Q Q Q D D D D N N N N P P P

6. 180   **7.** 55

Home Link 10·3

1. $0.06; $0.50; $1.30; $1.50; $3.36

3. 303   **4.** 197

Home Link 10·4

1. 1.09; 2.5; 0.98; 3.18; 0.06

3. 76   **4.** 72   **5.** 44   **6.** 18

Home Link 10·5

1. $0.70   **2.** $2.60   **3.** $1.00

4. $1.30   **5.** $4.00   **6.** $1.20

7. $2.30   **8.** $1.30 + $0.50 = $1.80

9. $0.80 + $0.40 = $1.20

10. $0.70 + $0.90 = $1.60

11. $1.40 + $0.80 = $2.20

Home Link 10·7

1. 17 sq cm   **2.** 23 cm^2 **3.** 11 square cm

4. 9 cm^2 **5.** 85   **6.** 29

Home Link 10·8

1. ④62   **2.** 1,③26   **3.** 5,⓪06   **4.** ⑧69

5. 2,③04   **6.** 4,⑤67   **9.** 1,183   **10.** 1,204

11. 158   **12.** 188   **13.** 29

Home Link 10·9

1. 0; 100; 200; 300; 400; 500; 600; 700; 800; 900; 1,000

2. 0; 1,000; 2,000; 3,000; 4,000; 5,000; 6,000; 7,000; 8,000; 9,000; 10,000

3.

Number	10 More	100 More	1,000 More
32	42	132	1,032
146	156	246	1,146
309	319	409	1,309
1,468	1,478	1,568	2,468
10,037	10,047	10,137	11,037

Home Link 10·10

3. 72,469   **4.** 72,569; 75,469; 72,369; 69,469

5. 76   **6.** 49   **7.** 225   **8.** 170

Home Link 10·11

1. 9   **2.** 15   **3.** 13   **4.** 6

5. 13 − (9 + 2) = 2

6. (28 − 8) − 4 = 16

7. (150 − 70) − 40 = 40

8. 800 − (200 + 300) = 300

9. **15**

~~25 = (15 + 5)~~
(25 − 15) + 5
(17 − 9) + 7
~~17 = (9 + 7)~~
(3 + 6) + 6
3 + (6 + 6)

10. **100**

(50 + 150) − 100
50 + (150 − 100)
~~400 = (300 = 200)~~
~~(400 − 300) + 200~~

HOME LINK 10·1 | **Coin Combinations**

Family Note In today's lesson, your child practiced writing amounts of money. For example, in Problem 1, 10 pennies can be written as 10¢ or $0.10. Your child also showed different groups of coins that have the same monetary value. For example, your child could show 62¢ with 2 quarters, 1 dime, and 2 pennies; or 4 dimes, 4 nickels, and 2 pennies. For Problem 2, help your child find items in newspaper or magazine ads and think of different combinations of coins and bills to pay for the items.

Please return this Home Link to school tomorrow.

1. Pretend that you have 10 of each kind of coin.
How much is that in all?

10 pennies = _____

10 nickels = _____

10 dimes = _____

10 quarters = _____

10 half-dollars = _____

Total = _____

2. Find two ads in a newspaper or magazine for items that cost less than $3.00 each.

   ◆ Ask for permission to cut out the ads.

   ◆ Cut them out and glue them onto the back of this page.

   ◆ Draw coins to show the cost of each item.

(If you can't find ads, draw pictures of items and prices on the back of this page.)

HOME LINK 10·2 | # How Much?

Family Note   In today's lesson, your child practiced reading and writing money amounts using dollars and cents. Ask your child to read each amount aloud. Remind your child that the digits before the decimal point stand for whole dollars; the digits after the decimal point stand for cents. When reading amounts such as "3 dollars and fifty-seven cents," the word "and" is used to denote the decimal point.

Please return this Home Link to school tomorrow.

MRB 90

How much money? Write your answer in dollars-and-cents notation.

1. \$1 \$1 \$1 Ⓠ Ⓠ Ⓝ Ⓟ Ⓟ $____._____

2. \$1 \$1 Ⓠ Ⓠ Ⓠ Ⓠ Ⓓ Ⓓ Ⓓ Ⓝ Ⓝ Ⓝ Ⓝ Ⓝ $____._____

3. Ⓠ Ⓓ Ⓓ Ⓟ Ⓟ Ⓟ Ⓟ Ⓟ Ⓟ Ⓟ $____._____

4. Ⓝ Ⓟ Ⓟ Ⓟ $____._____

5. Use \$1, Ⓠ, Ⓓ, Ⓝ, and Ⓟ to draw $2.64 in two different ways.

Practice

Solve.

6. 123 + 57 = _____   **7.** 84 − 29 = _____

HOME LINK 10·3 | Coin Values

Family Note In today's lesson, your child used a calculator to enter amounts of money and find totals. For Problem 2, help your child collect and find the total value of each type of coin. Then find the grand total. If you wish to use a calculator, help your child enter the amounts. Remind your child that amounts like $1.00 and $0.50 will be displayed on the calculator as "1." and "0.5" because the calculator doesn't display ending zeros.

Please return this Home Link to school tomorrow.

1. Complete the table.

Coins	Number of Coins	Total Value
P	6	$___.___
N	10	$___.___
D	13	$___.___
Q	6	$___.___
Grand Total		$___.___

2. Ask someone at home to help you collect pennies, nickels, dimes, quarters, and, if possible, half-dollars. Use the coins in your collection to complete the table below.

Coins	Number of Coins	Total Value
P		
N		
D		
Q		
Half-dollar		
Grand Total		

Practice

Solve.

3. 250 + 53 = _____

4. 250 − 53 = _____

HOME LINK 10·4 | Calculators and Money

Family Note
In today's lesson, your child used a calculator to solve problems with money. In Problem 2, your child will ask you or another adult to compare the cost of an item when you were a child to its current cost. There are two ways to make this type of comparison. You might describe a *difference comparison*. For example: "A bicycle costs about $90.00 more now than it did then." You might also use a *ratio comparison*. For example, "A bicycle costs about 4 times as much now as it did then." You do not need to share the terms *difference comparison* and *ratio comparison* with your child, but it is important that your child be exposed to both types of comparisons.

Please return this Home Link to school tomorrow.

1. Enter the following amounts into your calculator. What does your calculator show?

Enter	Calculator Shows
$1.09	_____
$2.50	_____
98¢	_____
$3.18	_____
6¢	_____

2. Ask an adult to think about an item that he or she remembers from when he or she was a child. Ask the adult to compare how much the item cost then and now. Record what you find out.

_____

_____

Practice

Solve.

3. 37 + 39 = _____

4. 49 + 23 = _____

5. 73 − 29 = _____

6. 56 − 38 = _____

HOME LINK 10·5 — Estimation to the Nearest 10¢

Family Note In today's lesson, your child estimated sums by first finding the nearest ten cents for each amount of money being added and then adding the amounts for the nearest ten cents together. For Problems 1–7, ask your child how she or he arrived at each answer. If needed, use coins to show which amount is actually closer. For Problems 8–11, help your child find the totals by thinking of a problem like $1.20 + $0.60 as 12 + 6 or as 120 cents + 60 cents.

Please return this Home Link to school tomorrow.

Write the correct answer to each question.
Talk with someone at home about your answers.

1. Is $0.69 closer to $0.60 or $0.70? _____

2. Is $2.59 closer to $2.50 or $2.60? _____

3. Is $0.99 closer to $0.90 or $1.00? _____

4. Is $1.31 closer to $1.30 or $1.40? _____

5. Is $3.99 closer to $3.90 or $4.00? _____

6. Is $1.17 closer to $1.10 or $1.20? _____

7. Is $2.34 closer to $2.30 or $2.40? _____

Fill in the blanks and estimate the total cost in each problem.

Example:

$1.19 + $0.59 is about *$1.20* + *$0.60* = *$1.80* .

8. $1.29 + $0.48 is about _____ + _____ = _____ .

9. $0.79 + $0.39 is about _____ + _____ = _____ .

10. $0.69 + $0.89 is about _____ + _____ = _____ .

11. $1.41 + $0.77 is about _____ + _____ = _____ .

HOME LINK 10·6 — Making Change

Family Note

In today's lesson, your child made change by counting up. When counting out change, encourage your child to begin with the cost of the item and count up to the amount of money that the customer has given to the clerk. For the example listed in the table below, your child could do the following:

1. Say "89 cents"—the price of the item.

2. Put a penny on the table and say "90 cents."

3. Put a dime on the table and say "$1.00."

4. Count the coins on the table. 1¢ + 10¢ = 11¢. The change is 11¢.

Please return this Home Link to school tomorrow.

Materials ☐ coins and bills (You can make bills out of paper.)

☐ items with prices marked

Practice making change with someone at home. Pretend you are the clerk at a store and the other person is a customer. The customer buys one of the items and pays with a bill. You count out the change.

Record some purchases here.

Item	Price	Amount Used to Pay	Change
can of black beans	$0.89	$1.00	$0.11

If possible, go to the store with someone. Buy something and get change. Count the change. Is it correct?

Name _____ Date _____ Time _____

Area

Family Note In today's lesson, your child found the area of shapes by counting square centimeters. As you observe your child finding the areas below, check that he or she is counting squares that are more than $\frac{1}{2}$ shaded as 1 square centimeter and not counting squares that are less than $\frac{1}{2}$ shaded. For Problem 4, see if your child has a suggestion for what to do if exactly $\frac{1}{2}$ of a square is shaded. Remind your child that area is reported in square units. Other ways to write square centimeters are **sq cm** and **cm²**.

Please return this Home Link to school tomorrow.

MRB 69

Count squares to find the area of each shaded figure.

1.

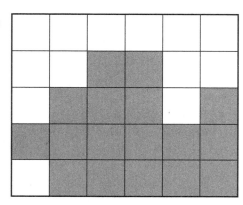

_____ sq cm

2.

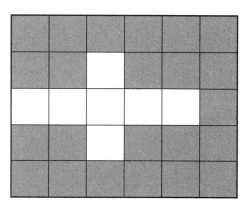

_____ cm²

3.

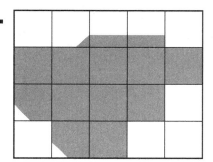

_____ square cm

4.

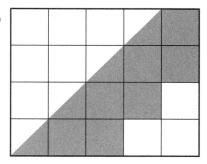

_____ cm²

Practice

5. 56
      + 29

6. 88
      − 59

211

HOME LINK 10·8 Place Value

Family Note　In this lesson, your child has been studying place value, or the value of digits in numbers. Listen as your child reads the numbers in Problems 1–6. You might ask your child to pick a few of the numbers and tell you the place value of each of the digits. For example, in 462, the value of 4 is 400, the value of 6 is 60, and the value of 2 is 2.

Please return this Home Link to school tomorrow.

MRB 10

In each number: ◆ Circle the digit in the hundreds place.

　　　　　　　　◆ Underline the digit in the thousands place.

Example: 9,③4 2

1. 4 6 2　　　　　**2.** 1 , 3 2 6　　　　**3.** 5 , 0 0 6

4. 8 6 9　　　　　**5.** 2 , 3 0 4　　　　**6.** 4 , 5 6 7

7. Read the numbers in Problems 1–6 to someone at home.

Write the numbers represented by the base-10 blocks.

8. 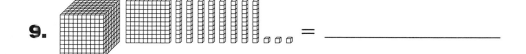 = _____247_____

9.  = _____

10. = _____

Practice

Solve.

11. $134 + 24 =$ _____　　　　**12.** $152 + 36 =$ _____

13. $67 - 38 =$ _____

HOME LINK 10·9 Counting by 10s, 100s, and 1,000s

Family Note In this lesson, your child used place value to count by 10s, 100s, and 1,000s. For Problems 1 and 2, listen carefully to find out if your child counts quickly and accurately. Help your child complete the table in Problem 3. If necessary, have your child use a calculator to find the answers. Ask your child to describe any patterns he or she sees in the completed table.

Please return this Home Link to school tomorrow.

MRB 162 163

1. Show someone at home how to count by 100s from 0 to 1,000. Record your counts.

_____

_____

2. Now count by 1,000s from 0 to 10,000. Record your counts.

_____

_____

3. Complete the table.

Number	10 More	100 More	1,000 More
32	42	132	1,032
146			
309			
1,468			
Try This			
10,037			

215

HOME LINK 10·10 | # 4-Digit and 5-Digit Numbers

Family Note In this lesson, your child read and displayed 4- and 5-digit numbers. Listen to your child read numbers to you. Remind your child not to say "and" when reading numbers such as the ones below. (In reading numbers, "and" indicates a decimal point. For example, 7.9 is read as "seven and nine tenths.") However, do not overcorrect your child if he or she inserts "and" occasionally.

Please return this Home Link to school tomorrow.

---

1. Read these numbers to someone at home.

   3,426;   6,001;   9,864;   13,400;   29,368;   99,999

2. Write other 4- and 5-digit numbers. Read your numbers to someone at home. _____

Try This

3. Write a number that has:

   4 in the hundreds place.

   6 in the tens place.

   2 in the thousands place.

   7 in the ten-thousands place.

   9 in the ones place.

   ___ ___ , ___ ___ ___

4. Use the number in Problem 3.

   What number is

   100 more? _____

   3,000 more? _____

   100 less? _____

   3,000 less? _____

Practice

5. 24 + 52 = _____

6. 78 − 29 = _____

7. 136
   + 89
   ———

8. 244
   − 74
   ———

HOME LINK 10·11 Grouping with Parentheses

Family Note In this lesson, your child has solved problems and puzzles involving parentheses. For Problems 1–4, 9, and 10, remind your child that the calculations inside the parentheses need to be done first. In Problem 1, for example, your child should first find $7 - 2$ and then add that answer (5) to 4. For Problems 5–8, observe as your child adds parentheses. Ask your child to explain what to do first to obtain the number on the right side of the equal sign.

Please return this Home Link to school tomorrow.

Solve problems containing parentheses.

1. $4 + (7 - 2) =$ _____

2. $(9 + 21) - 15 =$ _____

3. $6 + (12 - 5) =$ _____

4. $(15 + 5) - 14 =$ _____

Put in parentheses to solve the puzzles.

5. $13 - 9 + 2 = 2$

6. $28 - 8 - 4 = 16$

7. $150 - 70 - 40 = 40$

8. $800 - 200 + 300 = 300$

Cross out the names that don't belong in the name-collection boxes.

9.
15
$25 - (15 + 5)$
$(25 - 15) + 5$
$(17 - 9) + 7$
$17 - (9 + 7)$
$(3 + 6) + 6$
$3 + (6 + 6)$

10.
100
$(50 + 150) - 100$
$50 + (150 - 100)$
$400 - (300 - 200)$
$(400 - 300) + 200$

Unit 11: Family Letter

Whole-Number Operations Revisited

In the beginning of Unit 11, children will solve addition and subtraction stories with dollars and cents. Children will use estimation to examine their answers and determine whether the answers make sense.

Children will also review the uses of multiplication and division and begin to develop multiplication and division fact power, or the ability to automatically recall the basic multiplication and division facts.

Children will work with shortcuts, which will help them extend known facts to related facts. For example, the **turn-around rule for multiplication** shows that the order of the numbers being multiplied (the factors) does not affect the product; 3×4 is the same as 4×3. Children will also learn what it means to multiply a number by 0 and by 1. Working with patterns in a Facts Table and in fact families will also help children explore ways of learning multiplication and division facts.

×, ÷	1	2	3	4	5	6	7	8	9	10
1	1	2	3	4	5	6	7	8	9	10
2	2	4	6	8	10	12	14	16	18	20
3	3	6	9	12	15	18	21	24	27	30
4	4	8	12	16	20	24	28	32	36	40
5	5	10	15	20	25	30	35	40	45	50
6	6	12	18	24	30	36	42	48	54	60
7	7	14	21	28	35	42	49	56	63	70
8	8	16	24	32	40	48	56	64	72	80
9	9	18	27	36	45	54	63	72	81	90
10	10	20	30	40	50	60	70	80	90	100

Multiplication/Division Facts Table

Please keep this Family Letter for reference as your child works through Unit 11.

Vocabulary

Important terms in Unit 11:

multiplication diagram A diagram used in *Everyday Mathematics* to model situations in which a total number is made up of equal-sized groups. The diagram contains a number of groups, a number in each group, and a total number.

rows	_____ per row	_____ in all

Number model: _____ × _____ = _____

factor Each of the two or more numbers in a product.

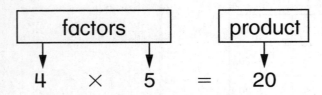

product The result of multiplying two numbers, called *factors*.

quotient The result of dividing one number by another.

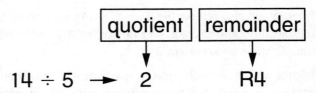

turn-around rule A rule for solving addition and multiplication problems saying it doesn't matter in which order the numbers are written. For example, if you know that $6 + 8 = 14$, then, by the turn-around rule, you also know that $8 + 6 = 14$.

range The difference between the largest (maximum) and smallest (minimum) numbers in a set of data. For example, the range of the data below is $38 - 32 = 6$.

32   32   34   35   35   37   38

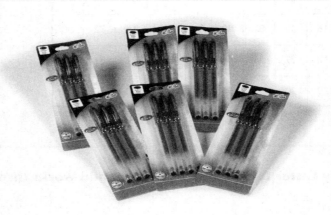

Do-Anytime Activities

To work with your child on the concepts taught in this unit and in previous units, try these interesting and rewarding activities:

1. Review common multiplication shortcuts. Ask, for example: *What happens when you multiply a number by 1? By 0? By 10?* Use pennies to show that 2×3 pennies is the same as 3×2 pennies.

2. At a restaurant or while grocery shopping, work together to estimate the bill.

3. Take turns making up multiplication and division number stories to solve.

Building Skills through Games

In Unit 11, your child will practice multiplication skills, mental arithmetic, and predicting the outcome of events by playing the following games:

Beat the Calculator

A "Calculator" (a player who uses a calculator to solve the problem) and a "Brain" (a player who solves the problem without a calculator) race to see who will be first to solve multiplication problems.

Hit the Target

Players choose a 2-digit multiple of ten as a "target number." One player enters a "starting number" into a calculator and tries to change the starting number to the target number by adding a number to it on the calculator. Children practice finding differences between 2-digit numbers and higher multiples of tens.

Array Bingo

Players roll the dice and find an *Array Bingo* card with the same number of dots. Players then turn that card over. The first player to have a row, column, or diagonal of facedown cards, calls out "Bingo!" and wins the game.

Name That Number

Each player turns over a card to find a number that must be renamed using any combination of five faceup cards.

Soccer Spin

Players pick which spinner will best help them make a goal.

As You Help Your Child with Homework

As your child brings home assignments, you may want to go over the instructions together, clarifying them as necessary. The answers listed below will guide you through this unit's Home Links.

Home Link 11·1

1. $2.22   **2.** $4.06   **3.** $3.34   **4.** $1.64

Home Link 11·2

1. glue stick; $0.14   **2.** glitter; $0.58

3. coloring pencils; $1.12

4. coloring pencils; $1.84

5. $0.11   **6.** $2.22

Home Link 11·3

1. 31; $70 - 40 = 30$ **2.** 23; $50 - 30 = 20$

3. 29; $90 - 60 = 30$ **4.** 17; $30 - 10 = 20$

5. 16; $30 - 20 = 10$

Home Link 11·4

1. 18 tennis balls; $6 \times 3 = 18$

2. 32 buns; $4 \times 8 = 32$

Home Link 11·5

1. 6 packages; $18 \div 3 \rightarrow 6$ R0

2. 6 cards; $25 \div 4 \rightarrow 6$ R1

Home Link 11·6

1. 12   **2.** 12   **3.** 10

4. 9   • • • • • • • • •

5. 14   • • • • • • •
           • • • • • • •

6. 12   • • • •
           • • • •
           • • • •

7. 2 nickels = 10 cents; $2 \times 5 = 10$

   6 nickels = 30 cents; $6 \times 5 = 30$

8. 4 dimes = 40 cents; $4 \times 10 = 40$

   7 dimes = 70 cents; $7 \times 10 = 70$

9. double 6 = 12; $2 \times 6 = 12$

   double 9 = 18; $2 \times 9 = 18$

Home Link 11·7

2. a. 99   **b.** 502   **c.** 0   **d.** 0

4. 55   **5.** 26

Home Link 11·9

1. $5 \times 7 = 35$ **2.** $3 \times 6 = 18$

   $7 \times 5 = 35$ $6 \times 3 = 18$

   $35 \div 5 = 7$ $18 \div 3 = 6$

   $35 \div 7 = 5$ $18 \div 6 = 3$

3. $4 \times 6 = 24$ **4.** $5 \times 6 = 30$

   $6 \times 4 = 24$ $6 \times 5 = 30$

   $24 \div 4 = 6$ $30 \div 5 = 6$

   $24 \div 6 = 4$ $30 \div 6 = 5$

HOME LINK 11·1 | **Buying Art Supplies**

> **Family Note**
>
> In today's lesson, your child solved number stories involving money amounts. Ask your child to explain to you how he or she solved each of the addition problems below. Challenge your child to find the total cost of 3 or 4 items. Encourage your child to use estimation before solving each problem. Ask such questions as: *Is the total cost of the crayons and glitter more or less than $3.00?* (less)
>
> *Please return this Home Link to school tomorrow.*

Crayons	**Glitter**	**Coloring Pencils**	**Glue Stick**
$0.75	$1.47	$2.59	$0.89

Find the total cost of each pair of items.

1. crayons and glitter Total cost: _____	**2.** glitter and coloring pencils Total cost: _____
3. crayons and coloring pencils Total cost: _____	**4.** glue stick and crayons Total cost: _____

225

HOME LINK 11·2

Comparing Costs

Crayons	Glitter	Coloring Pencils	Glue Stick
$0.75	$1.47	$2.59	$0.89

In Problems 1–4, circle the item that costs more.
Then find how much more.

1. glue stick or crayons

How much more? _____

2. glue stick or glitter

How much more? _____

3. glitter or coloring pencils

How much more? _____

4. coloring pencils or crayons

How much more? _____

5. Carlos bought a glue stick. He paid with a $1 bill. How much change should he get?

_____

6. Solve.

$1.47 + $0.75 = _____

227

HOME LINK 11·3 Trade-First Subtraction

Family Note

Today your child learned about subtracting multidigit numbers using a procedure called the trade-first method. Your child also used "ballpark estimates" to determine whether his or her answers made sense.

The **trade-first** method is similar to the traditional subtraction method that you may be familiar with. However, all the "regrouping" or "borrowing" is done before the problem is solved—which gives the method its name, "trade-first."

Example:

longs 10s	cubes 1s
4	6
− 3	9

◆ Are there enough tens and ones to remove exactly 3 tens and 9 ones from 46? *(No; there are enough tens, but there aren't enough ones.)*

◆ Trade 1 ten for 10 ones.

longs 10s	cubes 1s
3	1 6
4̶	6̶
− 3	9

◆ Solve. 3 tens minus 3 tens leaves 0 tens. 16 ones minus 9 ones leaves 7 ones. The answer is 7.

longs 10s	cubes 1s
3	1 6
4̶	6̶
− 3	9
	7

◆ Make a ballpark estimate to see whether the answer makes sense: 46 is close to 50, and 39 is close to 40. 50 − 40 = 10. 10 is close to the answer of 7, so 7 is a reasonable answer.

The trade-first method is one of many ways people solve subtraction problems. Your child may choose this method or a different procedure. What is most important is that your child can successfully solve subtraction problems using a method that makes sense to him or her.

*Please return the **second page** of this Home Link to school tomorrow.*

MRB 34

229

HOME LINK 11·3 **Trade-First Subtraction** *cont.*

Make a ballpark estimate for each problem and write a number model for your estimate.

Use the trade-first method of subtraction to solve each problem.

Example: Ballpark estimate:

$$30 - 20 = 10$$

longs 10s	cubes 1s
/1	/16
~~2~~	~~6~~
− 1	8
	8

Answer
8

1. Ballpark estimate:
_____

longs 10s	cubes 1s
7	3
− 4	2

Answer

2. Ballpark estimate:
_____

longs 10s	cubes 1s
4	9
− 2	6

Answer

3. Ballpark estimate:
_____

longs 10s	cubes 1s
8	5
− 5	6

Answer

4. Ballpark estimate:
_____

longs 10s	cubes 1s
3	2
− 1	5

Answer

5. Ballpark estimate:
_____

34 − 18

Answer

HOME LINK 11·4 | Multiplication Stories

Family Note

In today's lesson, your child solved multiplication number stories in which he or she found the total number of things in several equal groups. Observe the strategies your child uses to solve the problems below. The "multiplication diagram" is a device used to keep track of the information in a problem.

To solve Problem 1, your child would identify the known information by writing a 6 under *cans* and a 3 under *tennis balls per can*. To identify the unknown information, your child would write a ? under *tennis balls in all*.

Please return this Home Link to school tomorrow.

MRB
112 113

Show someone at home how to solve these multiplication stories.
Fill in each multiplication diagram.
Use counters or draw pictures or arrays to help you.

1. The store has 6 cans of tennis balls.
There are 3 balls in each can.
How many tennis balls are there in all?

cans	tennis balls per can	tennis balls in all

Answer: _____ tennis balls

Number model: _____ × _____ = _____

HOME LINK 11·4

Multiplication Stories *continued*

2. Hamburger buns come in packages of 8.
You buy 4 packages.
How many buns are there in all?

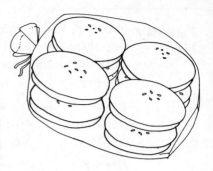

packages	buns per package	buns in all

Answer: _____ buns

Number model: _____ × _____ = _____

3. Make up and solve a multiplication number story below.

_____

_____

_____

_____	_____ per _____	_____ in all

Answer: _____

Number model: _____ × _____ = _____

HOME LINK 11·5

Division Number Stories

Family Note

Today your child solved division number stories about equal sharing and equal groups. The diagram used for multiplication can also be used for division number stories to identify known and unknown information. Your child will write a number model for each problem below. A number model is the symbolic representation of a number story. For example, in Problem 1, the number model is 18 ÷ 3 → 6 R0. This model is read as *18 divided by 3 gives 6, remainder 0*. An arrow is used instead of an equals (=) sign because the result of a division problem can be two whole numbers: the quotient and remainder.

Please return this Home Link to school tomorrow.

MRB
112–115

Show someone at home how to solve these division stories.
Use counters or draw pictures or diagrams to help you.

1. Our group needs 18 pens. There are 3 pens in each package. How many packages must we buy?

packages	pens per package	pens in all

Answer: _____ packages

Number model: _____ ÷ _____ → _____ R_____

2. Four children are playing a game with 25 cards. How many cards can the dealer give each player?

children	cards per child	cards in all

Answer: _____ cards

Number model: _____ ÷ _____ → _____ R_____

3. Make up and solve a division story on the back of this sheet.

HOME LINK 11·6 | **Multiplication Facts**

Family Note	In this lesson, your child has been learning multiplication facts and has used arrays to represent those facts. The first factor in a multiplication fact tells the number of rows in the array, and the second factor tells the number of columns in the array. In Problem 1, for example, an array with 2 rows of 6 dots is used for the multiplication fact $2 \times 6 = 12$.

Please return this Home Link to school tomorrow.

Show someone at home how you can use arrays to find products. Use •s.

1. $2 \times 6 =$ _____	**2.** $6 \times 2 =$ _____	**3.** $1 \times 10 =$ _____
• • • • • • • • • • • •		
4. $1 \times 9 =$ _____	**5.** $2 \times 7 =$ _____	**6.** $3 \times 4 =$ _____

7. 2 nickels = _____ cents $2 \times 5 =$ _____

6 nickels = _____ cents $6 \times 5 =$ _____

8. 4 dimes = _____ cents $4 \times 10 =$ _____

7 dimes = _____ cents $7 \times 10 =$ _____

9. double 6 = _____ $2 \times 6 =$ _____

double 9 = _____ $2 \times 9 =$ _____

HOME LINK 11·7 Multiplication Facts

Family Note In today's lesson, your child practiced multiplication facts by using a table and discussed patterns in multiplication facts. For example, any number multiplied by 1 is that number; any number multiplied by 0 is 0; and if the order of the factors in a multiplication fact is reversed, the product remains the same. Observe the strategies your child uses to find the answers below. Counting by 2s, 5s, 10s, and so on is one strategy to look for. Another strategy is drawing pictures. Some children may be able to solve some multiplication facts mentally, but this is not expected until the end of third grade.

Please return this Home Link to school tomorrow.

1. Show someone at home what you know about multiplication facts. You can use arrays or pictures to help solve the problems.

0 × 9 = _____	8 × 0 = _____	4 × 0 = _____	0 × 7 = _____
1 × 3 = _____	3 × 1 = _____	1 × 8 = _____	10 × 1 = _____
2 × 8 = _____	3 × 2 = _____	2 × 7 = _____	4 × 2 = _____
5 × 3 = _____	2 × 5 = _____	6 × 5 = _____	5 × 8 = _____
10 × 4 = _____	3 × 10 = _____	9 × 10 = _____	10 × 6 = _____

2. Explain to someone at home why it is easy to solve the following multiplication problems.

a. 99
$\times\ 1$

b. 502
$\times\ 1$

c. 37
$\times\ 0$

d. 15,461
$\times\ \ \ \ 0$

3. Make up and solve some multiplication problems of your own on the back of this page.

Practice

4. 84 − 29 = _____

5. 93 − 67 = _____

HOME LINK 11·8   ×, ÷ **Fact Triangles**

Family Note

Fact Triangles are tools for building mental arithmetic skills. You might think of them as the *Everyday Mathematics* version of the flash cards that you may remember from grade school. Fact Triangles, however, are more effective for helping children memorize facts because they emphasize fact families.

A **fact family** is a collection of related facts made from the same three numbers. For the numbers 4, 6, and 24, the multiplication/division fact family consists of $4 \times 6 = 24$, $6 \times 4 = 24$, $24 \div 6 = 4$, and $24 \div 4 = 6$.

Please help your child cut out the Fact Triangles attached to this letter.

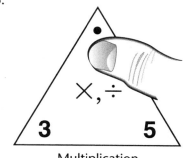

To use Fact Triangles to practice multiplication with your child, cover the number next to the dot with your thumb. The number you have covered is the product.

Multiplication

Your child uses the numbers that are showing to tell you one or two multiplication facts: $3 \times 5 = 15$ or $5 \times 3 = 15$.

To practice division, use your thumb to cover a number without a dot.

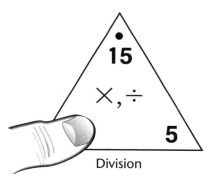

Your child uses the numbers that are showing to tell you the division fact $15 \div 5 = 3$.

Now cover the other number without a dot. Your child tells you the other division fact, $15 \div 3 = 5$.

Division

If your child misses a fact, flash the other two fact problems on the card and then return to the fact that was missed.

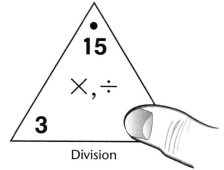

Example: Your child can't answer $15 \div 3$. Flash 3×5, then $15 \div 5$, and finally $15 \div 3$ a second time.

Make this activity brief and fun. Spend about 10 minutes each night. The work you do at home will support the work your child is doing at school.

Division

×, ÷ Fact Triangles *continued*

out the Fact Triangles. Show someone at home how you
use them to practice multiplication and division facts.

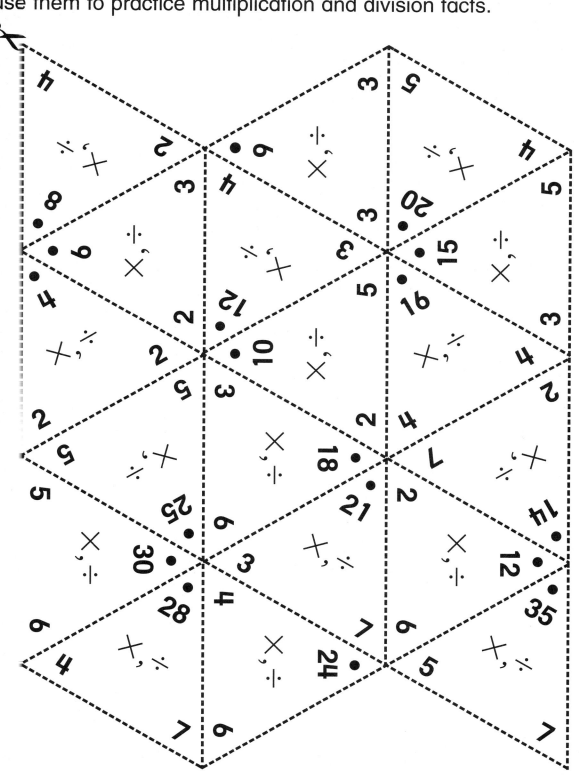

HOME LINK 11·9 Fact Families

Family Note Today your child continued to practice multiplication and division facts by playing a game called *Beat the Calculator* and by using Fact Triangles. Observe as your child writes the fact family for each Fact Triangle below. Use the Fact Triangles that your child brought home yesterday. Spend about 10 minutes practicing facts with your child. Make the activity brief and fun. The work you do at home will support the work your child is doing at school.

Please return this Home Link to school tomorrow.

MRB
38

Write the fact family for each Fact Triangle.

1.

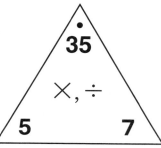

$\underline{5} \times \underline{7} = \underline{35}$

___ × ___ = ___

$\underline{35} \div \underline{5} = \underline{7}$

___ ÷ ___ = ___

2.

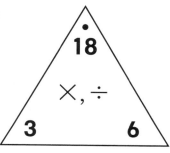

___ × ___ = ___

___ × ___ = ___

___ ÷ ___ = ___

___ ÷ ___ = ___

3.

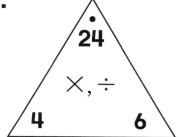

___ × ___ = ___

___ × ___ = ___

___ ÷ ___ = ___

___ ÷ ___ = ___

4.

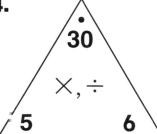

___ × ___ = ___   ___ ÷ ___ = ___

___ × ___ = ___   ___ ÷ ___ = ___

243

Unit 12: Family Letter

Year-End Reviews and Extensions

Rather than focusing on a single topic, Unit 12 reinforces some of the main topics covered in second grade.

Children will begin the unit by reviewing time measurements—telling time on clocks with hour and minute hands; naming time in different ways; using larger units of time, such as centuries and decades; and keeping track of time in years, months, weeks, and days.

Children will also work with computation dealing with multiplication facts and the relationship between multiplication and division.

Finally, children will display and interpret measurement data, with special attention to the range, median, and mode of sets of data.

Please keep this Family Letter for reference as your child works through Unit 12.

Vocabulary

Important terms in Unit 12:

timeline A *number line* showing when events took place. For example, the timeline below shows when the telephone, radio, and television were invented.

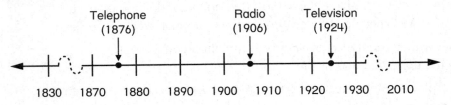

Telephone (1876)   Radio (1906)   Television (1924)

1830   1870   1880   1890   1900   1910   1920   1930   2010

mode The value or values that occurs most often in a set of data.

Building Skills through Games

In Unit 12, your child will practice adding and subtracting numbers by playing the following games:

Addition Card Draw

Each player draws the top 3 cards from a deck, records the numbers on the score sheet, and adds the 3 numbers. After 3 turns, players check each other's work with a calculator and add their 3 answers. The player with the higher total wins.

Name That Number

Each player turns over a card to find a number that must be renamed using any combination of five faceup cards.

Game 1

1st turn:

——— + ——— + ——— = ———

2nd turn:

——— + ——— + ——— = ———

3rd turn:

——— + ——— + ——— = ———

Total: ———

Do-Anytime Activities

To work with your child on the concepts taught in this unit and in previous units, try these interesting and rewarding activities:

1. Together, make up multidigit addition and subtraction number stories. Solve them. Share solution strategies.

2. Make timelines of your lives. In addition to personal information, mark various dates of events that interest you, such as events in music, art, sports, or politics.

3. Continue to ask the time. Encourage your child to name time in different ways, such as *twenty to nine* for 8:40 and *half-past two* for 2:30.

4. Continue to review and practice basic facts for all operations, emphasizing the multiplication facts.

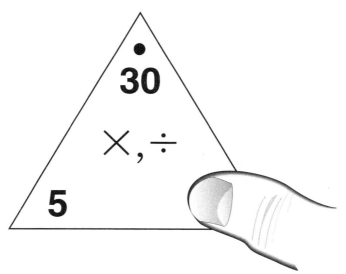

FEBRUARY						
Sun	Mon	Tue	Wed	Thu	Fri	Sat
	1	2	3	4	5	6
7	8	9	10	11	12	13
14	15	16	17	18	19	20
21	22	23	24	25	26	27
28	29					

As You Help Your Child with Homework

As your child brings home assignments, you may want to go over the instructions together, clarifying them as necessary. The answers listed below will guide you through this unit's Home Links.

Home Link 12◆1

1. 9 × 2 = 18
2 × 9 = 18
18 ÷ 2 = 9
18 ÷ 9 = 2

2. 1 × 8 = 8
8 × 1 = 8
8 ÷ 1 = 8
8 ÷ 8 = 1

3. 5 × 8 = 40
8 × 5 = 40
40 ÷ 8 = 5
40 ÷ 5 = 8

4. 184

5. 60

6. 243

7. 181

Home Link 12◆2

1. 4:10   **2.** 8:15   **3.** 10:45

4.

5.

6.

7. 169   **8.** 142   **9.** 91   **10.** 47

Home Link 12◆3

2. 531   **3.** 280

Home Link 12◆5

1. 7   **2.** 6   **3.** 7   **4.** 3
5. 4   **6.** 4   **7.** 4   **8.** 5
9. 7   **10.** 8   **11.** 7   **12.** 9
13. 6   **14.** 9

Home Link 12◆6

1. 30 years   **2.** dolphins and humans

3. 10 years   **4.** ostrich

5. squirrel, house cat, lion, horse, ostrich, dolphin, human

6. 30 years   **7.** 130   **8.** 156   **9.** 29   **10.** 87

Home Link 12◆7

1. a. 1,450   **b.** 1,750

2. a. 2,000   **b.** 1,300   **c.** 700

3. 1,450

4. 1,450

HOME LINK 12·1 | **Fact Triangles**

Family Note In class today, your child reviewed the calendar and continued to practice multiplication and division facts. Please spend a few minutes with your child as often as possible practicing facts. You can use Fact Triangles, or you can play a game like *Multiplication Top-It* or *Beat the Calculator.*

Please return this Home Link to school tomorrow.

MRB 38

Fll in the missing number in each Fact Triangle. Then write the fact family for the triangle.

1.

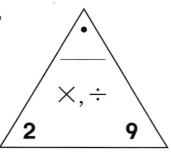

___ × ___ = ___

___ × ___ = ___

___ ÷ ___ = ___

___ ÷ ___ = ___

2.

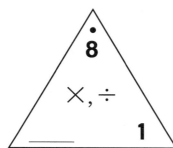

___ × ___ = ___

___ × ___ = ___

___ ÷ ___ = ___

___ ÷ ___ = ___

3.

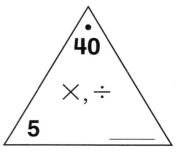

___ × ___ = ___

___ × ___ = ___

___ ÷ ___ = ___

___ ÷ ___ = ___

Practice

4.   231
       − 47

5.   85
       − 25

6.   156
       + 87

7.   94
       + 87

HOME LINK 12·2

Many Names for Times

Family Note Because clocks with clock faces were used for centuries before the invention of digital clocks, people often name the time by describing the positions of the hour and minute hands. Observe as your child solves the time problems below.

Please return this Home Link to school tomorrow.

82 83

What time is it? Write the time shown on the clocks.

1.

_____ : _____

2.

_____ : _____

3.

_____ : _____

Draw the hour hand and the minute hand to show the time.

4.

half-past nine

5.

six fifty

6.

quarter-to two

Practice

7. 126 + 43 =

8. 243 − 101 =

9. 38
      + 53
      ____

10.   84
        − 37
        ____

251

Timelines

HOME LINK 12·3

> **Family Note** A timeline is a way to display events in sequential order. Timelines can be divided into intervals, such as centuries, years, months, days, and hours. Observe your child as he or she completes the timeline at the right.
>
> *Please return this Home Link to school tomorrow.*

Emily's Day at the Beach

1. For each event below, make a dot on the timeline and write the letter for the event above the dot.

   A Ate lunch (12:30 P.M.)

   B Went fishing in a boat (10:00 A.M.)

   C Arrived at the beach (9:00 A.M.)

   D Returned from fishing trip (11:30 A.M.)

   E Played volleyball (1:30 P.M.)

   F Went swimming (2:00 P.M.)

   G Drove home (4:00 P.M.)

   H Built sandcastles (3:00 P.M.)

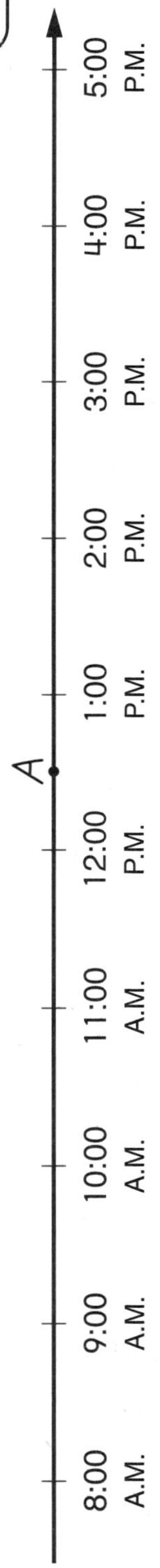

Practice

Solve.

2. 563 − 32 | **Answer** |
                | |

3.   263
      + 17 | **Answer** |
                | |

Timeline (right side):
5:00 P.M.
4:00 P.M.
3:00 P.M.
2:00 P.M.
1:00 P.M.
A ·
12:00 P.M.
11:00 A.M.
10:00 A.M.
9:00 A.M.
8:00 A.M.

253

HOME LINK 12·4 | **×, ÷ Fact Triangles**

Family Note Your child has been practicing multiplication facts. Today children reviewed shortcuts for solving multiplication problems with the numbers 2, 5, and 10. Encourage your child to practice with the Fact Triangles over the summer in preparation for third grade.

Cut out the Fact Triangles on these pages. Show someone at home how you can use them to practice multiplication facts.

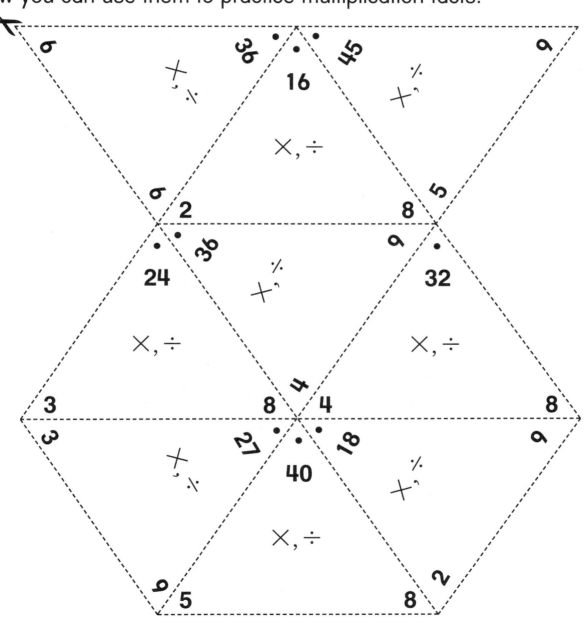

×, ÷ **Fact Triangles** *continued*

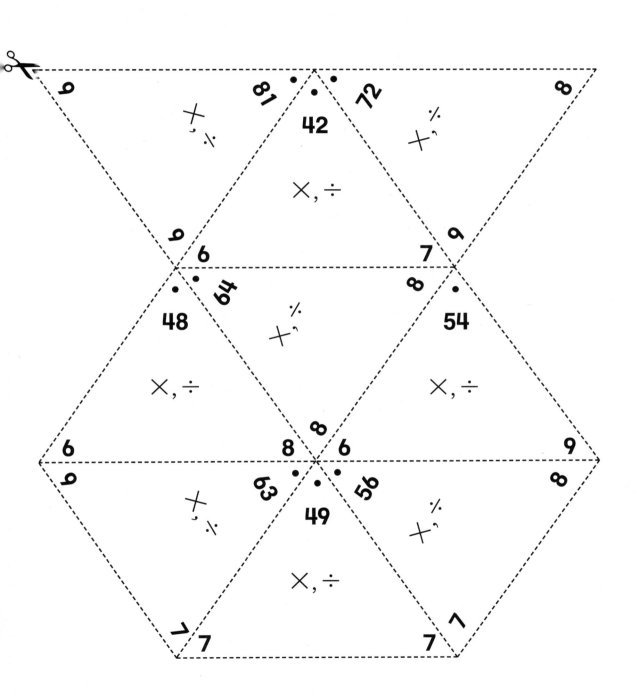

257

HOME LINK 12·5 | # ×, ÷ Facts Practice

> **Family Note** In this lesson, your child has connected multiplication and division facts by using Fact Triangles and completing fact families. A good way to solve division problems is to think in terms of multiplication. For example, to divide 20 by 5, ask yourself: *5 times what number equals 20?* Since 5 × 4 = 20, 20 ÷ 5 = 4.
>
> *Please return this Home Link to school tomorrow.*
>
> MRB 38

Solve these division facts. Think multiplication.

Use the Fact Triangles to help you.

1. 14 ÷ 2 = _____

Think:
2 × ? = 14

14
×, ÷
2   7

2. 24 ÷ 4 = _____

Think:
4 × ? = 24

24
×, ÷
4   6

3. 21 ÷ 3 = _____

Think:
3 × ? = 21

21
×, ÷
3   7

4. 18 ÷ 6 = _____

Think:
6 × ? = 18

18
×, ÷
6   3

5. 28 ÷ 7 = _____

Think:
7 × ? = 28

28
×, ÷
7   4

6. 16 ÷ 4 = _____

Think:
4 × ? = 16

16
×, ÷
4   4

×, ÷ **Facts Practice** *continued*

7. 20 ÷ 5 = _____

Think:
5 × ? = 20

20
×, ÷
5   4

8. 30 ÷ 6 = _____

Think:
6 × ? = 30

30
×, ÷
6   5

9. 35 ÷ 5 = _____

Think:
5 × ? = 35

35
×, ÷
5   7

10. 32 ÷ 4 = _____

Think:
4 × ? = 32

32
×, ÷
4   8

11. 42 ÷ 6 = _____

Think:
6 × ? = 42

42
×, ÷
6   7

12. 63 ÷ 7 = _____

Think:
7 × ? = 63

63
×, ÷
7   9

13. 54 ÷ 9 = _____

Think:
9 × ? = 54

54
×, ÷
6   9

14. 81 ÷ 9 = _____

Think:
9 × ? = 81

81
×, ÷
9   9

Typical Life Spans

Family Note In this lesson, your child has been reading, drawing, and interpreting bar graphs. Bar graphs are often useful when one wants to make rough comparisons quickly and easily. Provide your child with additional practice in interpreting a bar graph by asking questions like Problems 1 through 4.

Please return this Home Link to school tomorrow.

Typical Life Spans

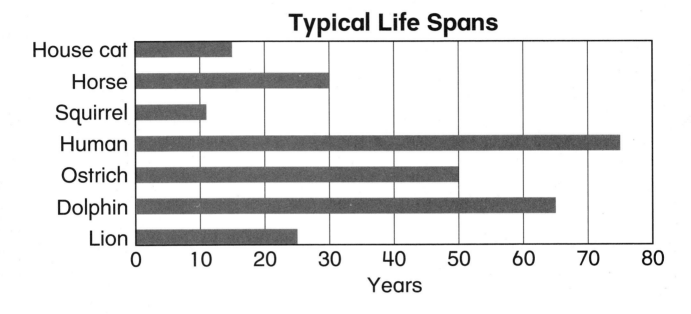

1. About how long do horses live? _____ years

2. Which animals live longer than an ostrich?

   _____

3. About how much longer do lions live than house cats?
   _____ years

4. Which animal lives about twice as long as lions? _____

HOME LINK
12·6

Typical Life Spans *continued*

5. List the animals in order from the shortest life span to the longest life span.

Life Spans	
Animal	**Years**
shortest:	
longest:	

6. What is the middle value? _____ years
This is the **median.**

| **Practice** |

7. 71 + 59 =

8.   121
   + 35
   ‾‾‾‾‾

9. 68 − 39 =

10.   125
   − 38
   ‾‾‾‾‾

262

HOME LINK 12·7   Interpret a Bar Graph

> **Family Note** In class today, your child interpreted graphs and identified the greatest value, the least value, the range, the middle value (the median), and the mode. The mode is the value or category that occurs most often in a set of data. For example, in the bar graph below, the river length of 1,450 miles is the mode.
>
> *Please return this Home Link to school tomorrow.*

Approximate Lengths of Rivers

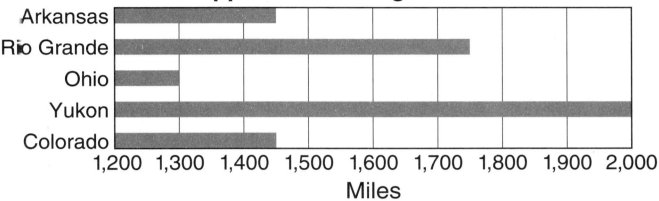

1. **a.** What is the length of the Colorado River? About _____ miles

   b. Of the Rio Grande? About _____ miles

2. **a.** What is the length of the longest river? About _____ miles

   b. What is the length of the shortest river? About _____ miles

   c. What is the difference in length between the longest and the shortest rivers? About _____ miles. This is the **range.**

3. Which river length occurs most often? About _____ miles This is the **mode.**

4. What is the middle length of the rivers? About _____ miles This is the **median.**

263

Family Letter

Congratulations!

By completing *Second Grade Everyday Mathematics,* your child has accomplished a great deal. Thank you for your support!

This Family Letter is provided as a resource for you to use throughout your child's vacation. It includes an extended list of Do-Anytime Activities, directions for games that can be played at home, an Addition/Subtraction Facts Table, and a sneak preview of what your child will be learning in *Third Grade Everyday Mathematics.* Enjoy your vacation!

Do-Anytime Activities

Mathematics concepts are more meaningful when they are rooted in real-life situations. To help your child review some of the concepts he or she has learned in second grade, we suggest the following activities for you and your child to do together over vacation. These activities will help your child build on the skills learned this year and help prepare him or her for *Third Grade Everyday Mathematics.*

1. Fill in blank calendar pages for the vacation months, including special events and dates. Discuss the number of weeks of vacation, days before school starts, and so on.

2. Continue to ask the time. Encourage alternate ways of naming time, such as *twenty to nine* for 8:40 and *quarter-past five* for 5:15.

3. Continue to review and practice basic facts for all operations, especially those for addition and subtraction.

4. Use Fact Triangle cards to practice basic multiplication and division facts, such as the following:

$2 \times 2 = 4$	$4 \div 2 = 2$
$2 \times 3 = 6$	$6 \div 2 = 3$
$2 \times 4 = 8$	$8 \div 2 = 4$
$2 \times 5 = 10$	$10 \div 2 = 5$
$3 \times 4 = 12$	$12 \div 3 = 4$
$3 \times 3 = 9$	$9 \div 3 = 3$
$4 \times 4 = 16$	$16 \div 4 = 4$
$3 \times 5 = 15$	$15 \div 3 = 5$
$4 \times 5 = 20$	$20 \div 4 = 5$

Building Skills through Games

The following section describes games that can be played at home. The number cards used in some games can be made from 3" by 5" index cards or from a regular playing-card deck.

Addition Top-It

Materials ☐ 4 cards for each of the numbers 0–10 (1 set for each player)

Players 2 or more

Skill Add, subtract, or multiply two numbers

Object of the Game To have the most cards

Directions

Players combine and shuffle their cards and place them in a deck, facedown. Each player turns up a pair of cards from the deck and says the sum of the numbers. The player with the greater sum takes all the cards that are in play. The player with the most cards at the end of the game is the winner. Ties are broken by drawing again—winner takes all.

Variation: *Subtraction Top-It*

Partners pool and shuffle their 0–20 number cards. Each player turns up a pair of cards from the facedown deck and says the difference between them. The player with the greater difference gets all four cards. The player with more cards at the end of the game is the winner.

Variation: *Multiplication Top-It*

Players find the product of the numbers instead of the sum or difference. Use the 0–10 number cards.

Pick-a-Coin

Materials
☐ regular die
☐ record sheet (see example)
☐ calculator

Players 2 or 3

Skill Add coin and dollar amounts

Sample Record Sheet						
	Ⓟ	Ⓝ	Ⓓ	Ⓠ	$1	Total
1st turn	2	1	4	5	3	$ 4.72
2nd turn						$ _._
3rd turn						$ _._
4th turn						$ _._
					Grand Total	$ _._

Object of the Game To have the highest total

Directions

Players take turns. At each turn, a player rolls a die five times. After each roll, the player records the number that comes up on the die in any one of the empty cells for that turn on his or her Record Sheet. Then the player finds the total amount and records it in the table.

After four turns, each player uses a calculator to find his or her grand total. The player with the highest grand total wins.

Multiplication Draw

Materials
☐ number cards 1, 2, 3, 4, 5, 10 (4 of each)

☐ record sheet (1 for each player)

☐ calculator

Players 2–4

Skill Multiply two numbers

Object of the Game To have the highest total

Multiplication Draw Record Sheet

1st Draw: _____ × _____ = _____

2nd Draw: _____ × _____ = _____

3rd Draw: _____ × _____ = _____

4th Draw: _____ × _____ = _____

5th Draw: _____ × _____ = _____

Sum of products: _____

Directions

Shuffle the cards and place the deck facedown on the playing surface. At each turn, players draw two cards from the deck to make up a multiplication problem. They record the problem on a record sheet and write the answer. If the answer is incorrect, it will not be counted. After five turns, players use a calculator to find the total of their correct answers. The player with the highest total wins.

Name That Number

Materials
☐ number cards 0–10 (4 of each)

☐ number cards 11–20 (1 of each)

Players 2 or 3

Skill Add, substract, multiply, or divide two numbers to reach a target number

Object of the Game To have the most cards

Directions

Shuffle the deck of cards and place it facedown on the table. Turn the top five cards faceup and place them in a row. Turn over the next card. This is the target number for the round.

In turn, players try to name the target number by adding, subtracting, multiplying, or dividing the numbers on 2 or more of the 5 cards that are number-side up. A card may be used only once for each turn. If you can name the target number, take the cards you used to name it. Also take the target-number card. Then replace all the cards you took by drawing from the top of the deck. If you cannot name the target number, your turn is over. Turn over the top card of the deck and lay it down on the target-number pile. The number on this card is the new target number.

Play continues until there are not enough cards left in the deck to replace the players' cards. The player who has taken the most cards at the end wins. Sample turn:

Mae's turn: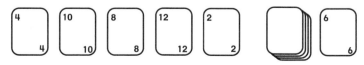

The target number is 6. Mae names it with 12 − 4 − 2. She also could have used 4 + 2 or 8 − 2.

Mae takes the 12, 4, 2, and 6 cards. She replaces them by drawing cards from the facedown deck and then turns over and lays down the next card to replace the 6. Now it is Mike's turn.

Fact Power

Addition/subtraction fact families can also be practiced by using the Addition/Subtraction Facts Table. This table can be used to keep a record of facts that have been learned as well.

+,−	0	1	2	3	4	5	6	7	8	9
0	0	1	2	3	4	5	6	7	8	9
1	1	2	3	4	5	6	7	8	9	10
2	2	3	4	5	6	7	8	9	10	11
3	3	4	5	6	7	8	9	10	11	12
4	4	5	6	7	8	9	10	11	12	13
5	5	6	7	8	9	10	11	12	13	14
6	6	7	8	9	10	11	12	13	14	15
7	7	8	9	10	11	12	13	14	15	16
8	8	9	10	11	12	13	14	15	16	17
9	9	10	11	12	13	14	15	16	17	18

Looking Ahead:
Third Grade Everyday Mathematics

Next year, your child will …

◆ Explore the relationship between multiplication and division

◆ Extend multiplication and division facts to multiples of 10, 100, and 1,000

◆ Use parentheses in writing number models

◆ Record equivalent units of length

◆ Use number models to find the areas of rectangles

◆ Explore 2- and 3-dimensional shapes and other geometric concepts

◆ Read and write numbers up to 1,000,000

◆ Work with fractions and decimals

◆ Collect data for yearlong sunrise/sunset and high/low temperature projects

◆ Use map scales to estimate distances

Again, thank you for your support this year. Have fun continuing your child's mathematics experiences throughout the vacation!